IPHONE SE USER GUIDE

FOR SENIORS

HOW TO USE YOUR NEW IPHONE SE CORRECTLY. THE BEGINNER'S GUIDE TO TEACH IN A SIMPLE AND PRECISE WAY WHAT HAS ALWAYS CONFUSED YOU SO FAR!

By: **BRANDON BALLARD**

Table of Contents

Introduction

Apple wanted to give users a new iPhone at a lower price, which is why they launched the 2020 iPhone SE. For just $399, you get to use a new Apple smartphone and still enjoy the great features that the Apple brand is associated with. For this price, you get to enjoy the blazing A13 Bionic processor, which gives you not just the camera capabilities but also the fastest performance that one will expect from a pricier phone—all this greatness wrapped in a compact design for lovers of small phones.

The new iPhone SE is similar to the iPhone 8 but has a lot of premium features that are not available on Android phones in this price range. These features include wireless charging, metal & glass design, not forgetting its water and dust resistance.

The 2020 iPhone SE was released on April 24 at the cost of $399/£419 for the 64 GB storage. However, users can get the 128GB storage for £469/ $449, while the 256 GB storage goes for $549/£569.

Display

The new iPhone SE has a screen size of 4.7 inches. This may be a small screen for some, but for others, the screen size is perfect. You will enjoy playing games or streaming videos online with your iPhone SE.

The device also has a bright screen with its rating of 625 nits. The phone has a display resolution of 1334 x 750 pixels, which is not a bad trade-off when compared with other phones in the same price line. The phone also supports True Tone, which makes it possible for the screen color to adjust automatically based on the lighting conditions.

Performance

The iPhone SE 2020 has the same A13 Bionic chip that the iPhone 11 and 11 Pro has, giving you the same fast and responsive performance. You can switch between apps or load apps in seconds.

Colors and Design

The iPhone SE has three color options: white, black, and red. The phone is also made with the same sturdy glass and aluminum design that is on the iPhone 8.

The phone has IP67 water resistance and so can be submerged in one meter of water for up to thirty minutes. However, Apple has warned that you should not intentionally expose the phone to water to avoid wear and tear that is common with any device or gadget.

Camera

Because the iPhone SE has the same processor as the iPhone 11 series, you may not be able to tell the difference between the photos captured on the iPhone SE and that of the iPhone 11. However, the iPhone SE does not have the Night Mode feature that is available on the iPhone 11 due to the single camera lens on the iPhone SE.

The iPhone SE has a 7MP front camera and a single 12-megapixel rear camera. It does not have a telephoto lens with an optical zoom like the iPhone 11 Pro or an ultra-wide-angle lens like the iPhone 11.

But the A13 Bionic processor helps the performance of the iPhone SE camera in numerous ways. For one, the iPhone SE camera enjoys the Smart HDR that brings out the highlights in faces. You can take portraits with the front and back camera. You also have the whole Portrait Lighting effects as well as the ability to customize the depth of field.

You can also record 4K videos at up to 60 fps. Change the frame rate to 30 fps and enjoy an extended dynamic range.

Touch ID

Apple, on all its newer phones, has always offered Face ID. However, the new iPhone SE uses a Touch ID sensor to unlock the device, authorize Apple Pay, App Store downloads, and entering of passwords. The Touch ID particularly comes useful at this time when we need to wear face masks in public. Face ID will not work when you wear a face mask, but the Touch ID does not require you to take off your mask to unlock your device.

Battery

The battery life of the iPhone SE 2020 is not the same as that of the iPhone 11 series. According to Apple, the battery life of the iPhone SE is almost the same as the iPhone 8, with its 1,821mAh power pack. Apple confirmed that the new iPhone SE gives 13 hours of video playback, while the iPhone 11 has 17 hours of video playback.

The phone also comes with the standard 5-watt Lighting charger, which gives you about 29 percent power in thirty minutes. However, you can get the 18W charger from Apple if you want to enjoy faster charging. With the fast charger, you can go from empty to over 50 percent in about 30 minutes. This 18W fast charger is sold for $29 for the power brick and another $19 if you want the USB C to Lightning cable.

Another good news is that the phone supports 7.5 wireless charging, as the 2020 iPhone SE is Qi-charging enabled. Get a cheap wireless pad from online stores or any gadget stores near you, and you will never experience issues with the power levels on your phone.

Apart from the few hitches here and there, the 2020 iPhone SE is still a great buy as you get to enjoy almost all the features that are available in the higher-priced iPhone 11, iPhone 11 Pro, and iPhone 11 Pro Max. Let us now explore all the features that the phone offers.

How to Turn On/Off the iPhone

- Go to **Settings > General** and scroll to the very bottom where you will see the **Shutdown option**.

- Click the **Shutdown option**. Drag the **Slide to Power Off** slider to the right to power off your iPhone.

- Alternatively, to get into the **Switch-off menu**, you must hold down one of the two-volume buttons in addition to the side button (formerly the power button), and you will see a **Slide to Power Off** option. Drag it to the right to switch off your device.

Set Up iPhone SE

Whether you're brand new to the Apple ecosystem or the tenth update, setting up a new iPhone is a fun experience, no different than waking up for Christmas morning. From when the first "Hello" appears to the last step, here's all you need to know when setting up your new device.

Understand Your Options

Use one of these three ways to set up your new iPhone, update, or restore from your old iPhone, transfer files from a non-Apple phone. Here are more detailed meanings of each of these options.

Setting as new means that everything—all settings—start from scratch. This is for people who have never used a smartphone or online services before or who want their iPhone to feel truly new.

Restore from a previous backup of iPhone, iPad, or iPod touch—You can do this online via iCloud or via USB via iTunes or Finder (macOS Catalina). This is for people who already have an iOS device and are switching to a new one who wants everything they have on the old device to be intact on the new one.

Import from Android, BlackBerry, or Windows Phone—Apple has an app on Google Play to make Android easier, but online services allow you to move a lot of data from any old device. This is for people who switch to iPhone or iPad from another mobile platform.

How to Start Setting Up a New iPhone

As soon as you turn on your new iPhone for the first time, you will be greeted by "Hello" in different languages. The same goes for whether you start from scratch, restore from another iPhone, or switch from Android.

- Tap the slide to customize and swipe the screen to get started.

- Choose your language.

- Select your country or region.

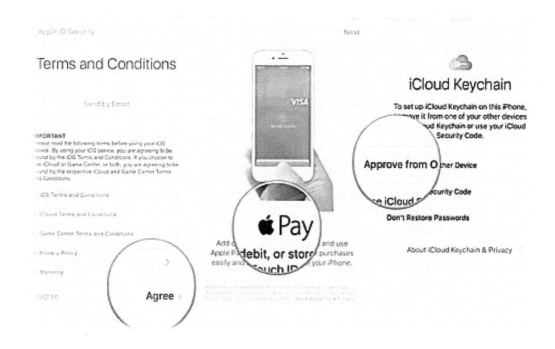

- Select a **Wi-Fi network**. If you don't have a Wi-Fi network, you can configure it later. Select **Cellular** instead.

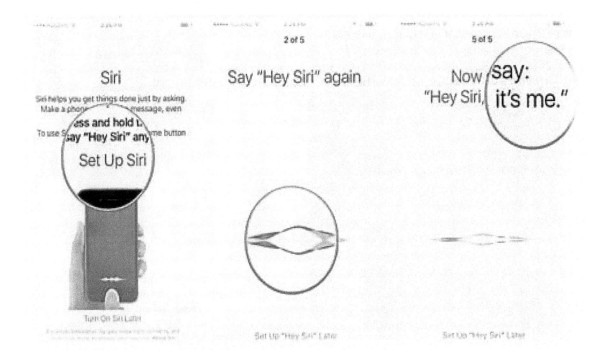

- At this point, you can use the automatic setup to set up your new iPhone with the same password and settings as other iPhones. If you decide to set up your new iPhone manually, follow these steps:

➢ Click **Continue** after reading the information about **Apple data and privacy**.

➢ Tap **Enable location services**. If you do not want to enable location services now, select **Skip location services**. You can turn on some location services manually, such as **Maps**.

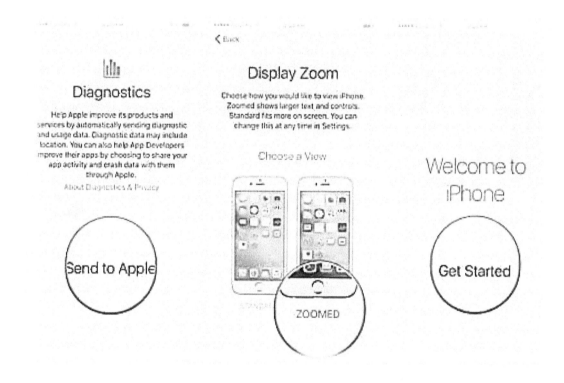

Set Up With Touch ID

iPhone SE (2020) has the same design as the iPhone 8 but takes the A13 bionic chip processing power of the iPhone 11 series. However, unlike the iPhone 11 series, the iPhone SE still uses the home screen button and Touch ID for security and biometrics. It's easy to set up, and you can find setup instructions here.

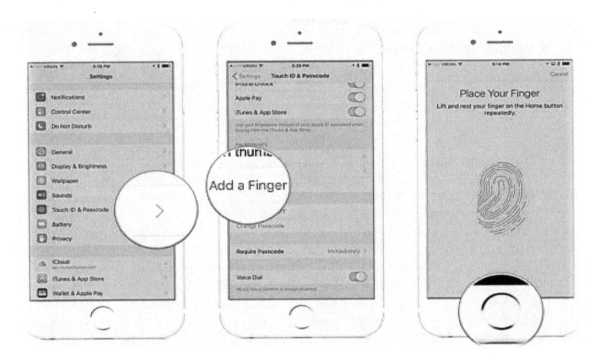

How to Use Touch ID

The iPhone X, XS, XS Max, XR, 11, 11 Pro, and 11 Pro Max no longer have a Home button, and Touch ID has been replaced by Face ID for security and biometrics. Set it similar to Touch ID, but you use a face instead of a thumb. You can find every step to customize this here.

How to Set up Face ID on a New iPhone

You will then be asked if you want to restore from a backup, install it as a new iPhone or move data from Android.

How to Recover or Transfer Data From Another Phone

If you don't start the upgrade with a new data cleaner, you'll either have to transfer data from your old iPhone to your new one or transfer data from your old Android device to your new iPhone. Here's how to do it.

How to Recover From iCloud or iTunes Backup

It's time to decide how you want to transfer your old iPhone data (if you're starting from scratch, learn how to set up the iPhone as new). You have two options for recovering apps and data from another iPhone; iCloud or iTunes/Finder.

Which one you choose depends on whether you've backed up your old iPhone to iCloud, connected it to your computer, and backed it up with iTunes or Finder.

The key here is to make sure that your old iPhone was backed up first.

After backing up your old iPhone, choose whether you want to restore your new iPhone from iCloud or iTunes.

How to Migrate Files From Old Android Device

If you're moving to iPhone from an Android-based operating system, welcome to the Apple family.

Apple has a special app for those who move with Android called Move to iOS, and it's available in the Google Play Store. Before transferring data to a new iPhone, download Move to iOS to your Android phone.

How to Set Up iPhone as New

If this is your first iPhone and you don't want to migrate Android data, or if this is your tenth iPhone and you just want to start over, you can set up your iPhone as new.

- Click **Set as new iPhone**.

- Enter your Apple ID and password or create a new one. Touch **Don't have an Apple ID?** and follow the instructions.

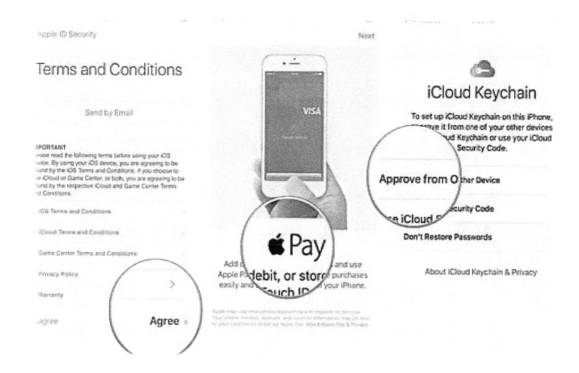

- Read and accept **Apple's terms of use**.

- Click **Agree** again to confirm.

- Set up **Apple Pay**.

- Set up an **iCloud keychain**.

- Agree to the terms, configure **Apple Pay**, and then configure the **iCloud keychain.**

- Set up **Siri** and **Hey, Siri.**

- Tap **Send Apple Diagnostics** in the event of a program crash or other problem, or tap **Don't send** if you don't want to.

- Turn on **Zoom Display** for greater visual accessibility.

- Click the **Start button**.

- Send or not send diagnostics to Apple, zoom in or out, and get started!

How to Use Internet and How to Activate the Connection

To connect your phone to Wi-Fi, go to your phone settings and click on **Wi-Fi**. Select the Wi-Fi network you want to connect to and click on it. Depending on the type of the Wi-Fi network, you may have to provide a password (for secure connections). Enter the password and click on **Join** to successfully connect to the Wi-Fi network. If you were able to connect, a blue tick would appear next to the Wi-Fi network.

How to Connect to the Internet

Besides Wi-Fi, you can also use your cellular network to connect your phone to the internet. To do this, go to your phone settings and click on **Cellular**. With your cellular data turned on, you can now connect to the internet. For your phone to be able to connect to the internet, you need to have a working sim card (nano-sim) and a data plan. You can contact your service provider for more information.

Turn on iCloud Backup

You may not always remember to backup your data. Apple knows this and makes it possible for your phone to automatically backup your data daily. To enable this feature, go to your phone settings. Once you are in your phone settings, click on your **Apple ID** and turn on your **iCloud**. You should ensure your iPhone is connected to Wi-Fi network and your phone is connected to a power source. You can still backup your data manually to iCloud using your mobile data. You can get up to 50GB to save your data on iCloud. You can purchase an extra 50GB for $0.99 a month.

How to Customize Assistive Touch

Assistive touch helps bypass a number of procedures. To enable assistive touch, you should go to your phone settings. Click on **Genera** and further click on **Accessibility**. Under the accessibility sub-menu, click on the **Physical and motor sub-menu**, and you will find assistive touch. You can turn it on and further customize it. It usually appears as a circle constantly on your phone screen. You can also change the position if it is obstructing how you use your phone by dragging the circle to wherever you want it to be.

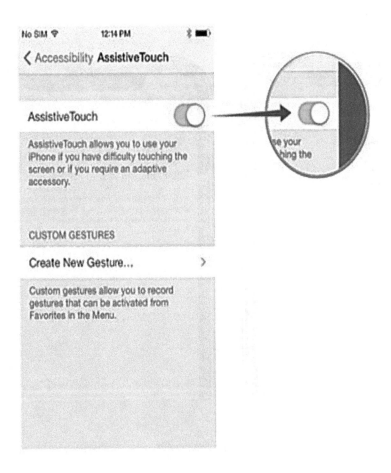

Set Messages to Share Your Personalized Contact Data

New to iOS 13 is the option to create your very own contact photo and name to be displayed on other people's iPhone device. You can pick whether this is enabled for just contacts or everyone; however, they have the last say on whether they acknowledge your chosen information.

Tap **Settings > Messages > Share Name & Photo** where you can configure these and whom this automatically gets shared with.

How to Scan Documents in the Files App

Open the **file app.**

On the navigation screen, tap the **three-dot button** (three-dot circle) in the upper right corner of the screen, besides the folder where you want to save the scans, then swipe down the screen to show the **Options Bar folder** and tap the **three-dot button** on the left.

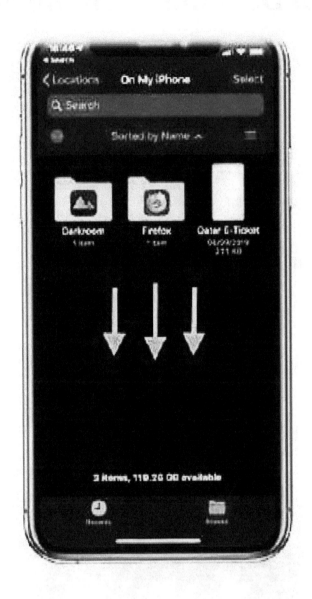

Select **Scan Documents** from the **drop-down menu**.

By default, the camera automatically tries to recognize and scan a document in the viewfinder. If you don't want this, tap **Automatic** in the upper right corner of the camera surface to switch to Manual.

At the top, tap the ellipses and choose **color**, **grayscale**, **black and white**, or **scan photos**. The default option is **color**. If you need to configure the flash options, tap the **flash icon.** The default setting is automatic. This turns the flash off when you are in a dimly lit room.

Make sure the yellow box is aligned with the edges of the document by focusing the camera on your document.

When you're aligned, tap the **camera's shutter button** to take a photo.

Adjust the edges of the scan for perfect alignment. The application automatically corrects all outstanding.

If you want to scan and don't want to scan multiple pages, tap **Done** and then tap **Save** when you return to the scanning interface; if you want to try the same scan again, select **Retry**.

If you started the scan using the **Browse app** via the **File app**, you will be asked to select a folder in which the scanned documents will be saved. Otherwise, the scanned documents are automatically saved as PDF in the previously examined folder.

Apple's document scanning tools are well designed and impressive. They deliver dozens of clear and clean scans in our tests, and they perform excellently for everything from photos to documents. Apple's scanning tool can also easily compete with and replace established third-party document scanners.

Display Multitasking Quick App Switcher

- Touch your finger to the gesture area at the bottom of the iPhone SE display.

- Swipe up slightly. (Try not to flick. Simply keep your finger on the screen until you get a short far up, then pull away.)

How to Access Reachability Mode

- Open **Settings** from the Home screen.

- Tap on **General**.

- Tap on **Accessibility**.

- Switch **Reachability** to **ON.**

How to Access Control Centre

- Touch your finger to the gesture area at the extreme bottom of the iPhone SE display.

- Swipe down.

- Once more, you can even swipe down from the top right of **Reachability** to access **Control Center**.

How to Create a New Apple Id and Set Up Apple Play

This is an Apple account that lets you download and install games and apps from the Apple store, purchase books, movies, and music from iTunes, sync your contacts, reminders, and calendars through iCloud as well as use iMessage and Facetime in the messages apps.

Create an Apple ID

- Open the Settings app.

- Navigate to the top of the screen and click on **Sign in to your iPhone**.

- Click on **Don't have an Apple ID or forgot it?**

- On the next screen, select **Create Apple ID**.

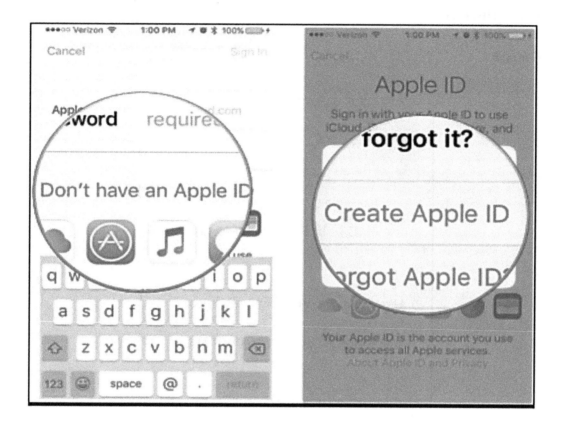

- Input your date of birth, then tap **Next**.

- Input your names: first and last, then tap **Next**.

- Choose to **Sign up with a current email address** or choose to **Get a free iCloud email address**.

- Input your email address and create your desired password.

- Repeat the password.

- Set a security question and input the answer. You are to select three security questions and input their answers.

- Click on **Next**.

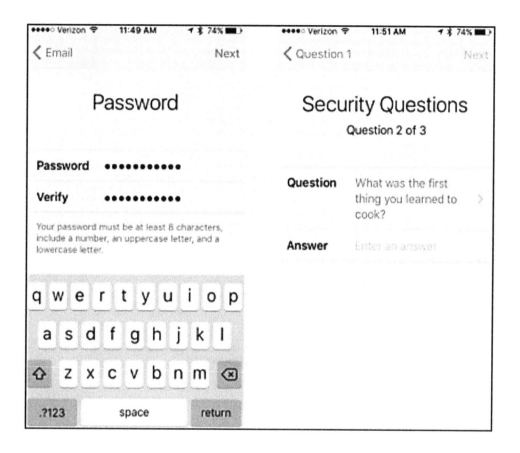

- Accept the **Terms and Conditions**.

- Choose to **Merge** or **Not to merge**, to sync iCloud data from calendars, contacts, Safari, and reminders.

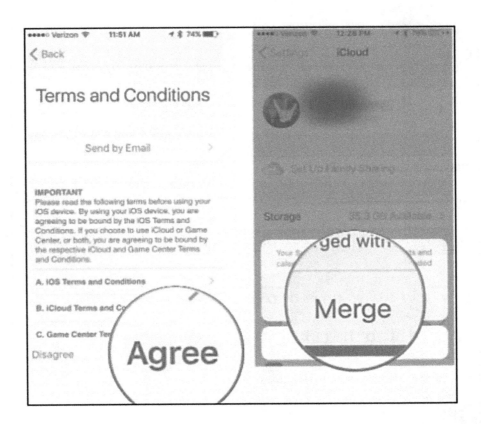

- Click on **Ok** to confirm that you want the **Find My Enabled**.

Sign in to iCloud With an Existing Apple ID

- Open the **Settings** app.

- Navigate to the top of the screen and click on **Sign in to your iPhone.**

- Input your login details, then click on **Sign In.**

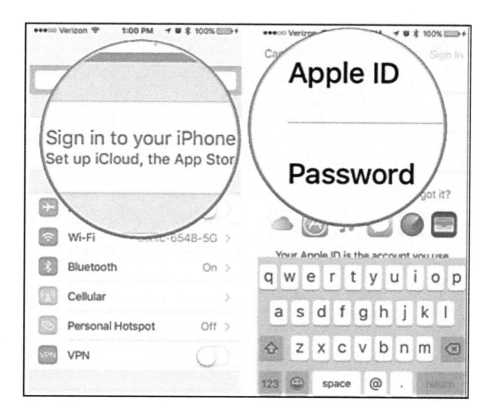

- Enter your passcode when prompted.

- Check that your iCloud photos are in the way you like them.

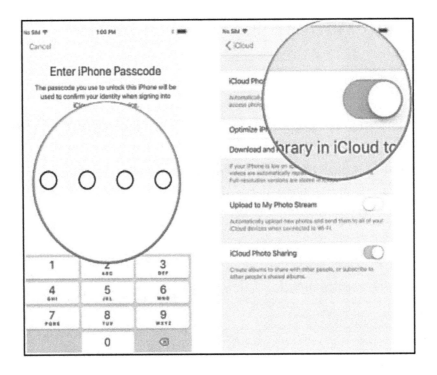

- Enable or disable the option for Apps using iCloud, depending on your preference.

Sign Out of iCloud

- Open the Settings app.

- Navigate to the top of the screen and click on your **Apple ID**.

- Go to the bottom of the screen, then click on **Sign Out.**

- Input your Apple ID password for this account, then click on **Turn Off.**

- Toggle the switch for all the data you want to keep on your device.

- Navigate to the top right side of the screen and click on **Sign Out.**

- Click on **Sign Out** again to confirm your action.

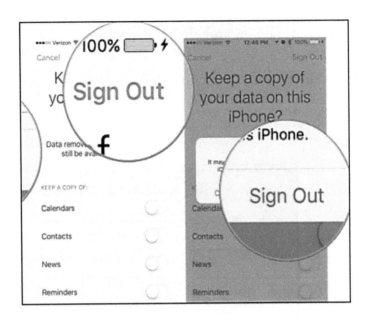

Apple Pay

Apple Pay allows you to make online and in-store purchases on your iPhone, just by tapping the home button and scanning your fingerprint. This is a fast and secure way to make purchases using your debit and credit cards.

Add a Card for Apple Pay

- Launch the wallet app on your phone, then tap the ⊕ button at the top side of your screen.

- Tap **Next**.

- Scan your card information or click on **Enter Card Details Manually** to input it manually.

- Click on **Next** on the details screen.

- Manually type the expiry date and security code of the card.

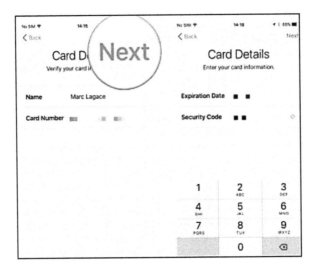

- Click **Next** and agree to the Terms and Conditions.

- Tap **Agree** again.

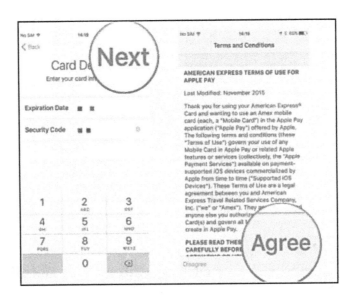

- Select your verification method and click on **Next**.

- Click on **Enter Code.**

- Input the verification code sent to you via text, call, or email.

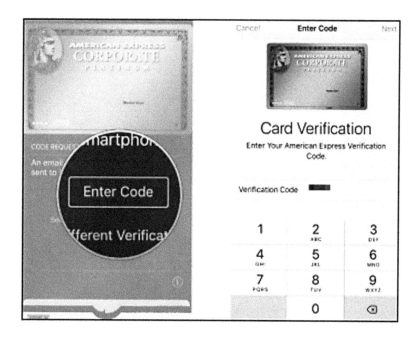

- Click on **Next**, then click on **Done**.

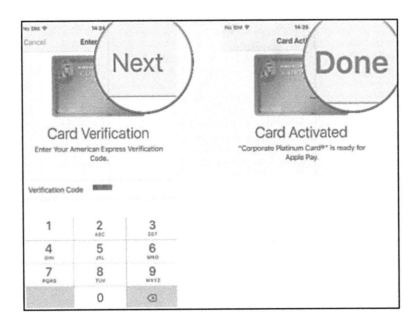

- Repeat the steps above to card more cards in the future.

Note: credit or debit cards with flat numbers will have to be inputted manually. Only cards with embossed numbers can be scanned into the Apple Pay wallet.

Change the Default Card for Apple Pay

While you can always switch between cards when performing transactions, using the default card makes your transaction fast and easy. Here is how to set a default card:

- Open the Settings app on your phone.

- Click on **Wallet & Apple Pay.**

- Then click on **Default Card.**

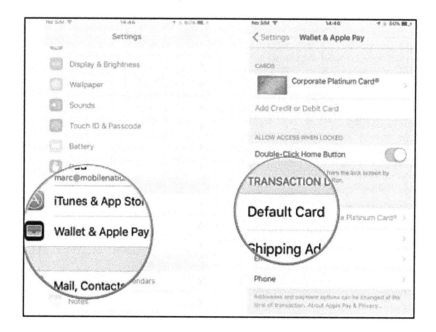

- Click on the card you want to set as your default.

- When next you want to make a purchase, your default card will be used for payment.

Remove a Card From Apple Pay

Here is how to remove a card when lost, stolen, or for other reasons:

- Open the Settings app on your phone.

- Click on **Wallet & Apple Pay.**

- Click on the card you wish to delete.

- Then scroll to the bottom and click on **Remove Card.**

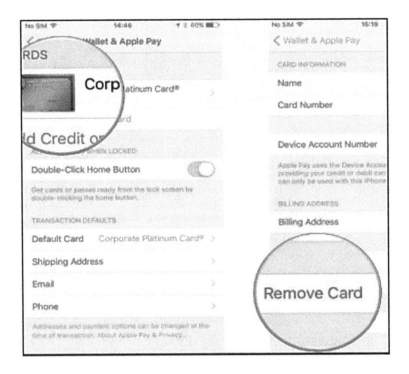

Note: this will only remove the card on your iPhone SE. You will have to manually remove the card on all other devices that have the card linked. An alternative is to remove the card remotely via iCloud. Below is how to do this.

How to Enable Emergency SOS

The emergency SOS comes in handy as your phone will automatically be able to call the emergency numbers in your country. It can even send them your location. To enable emergency SOS, go to your phone settings, click on **Emergency SOS**, and switch it on. The emergency SOS works with an audio countdown. You can turn off the audio countdown. Under the emergency SOS sub-menu, you will see a **Countdown sound** option. Toggle it off to remove the audio countdown that occurs during an emergency SOS call. You can also add contacts to your emergency using your health app. Open your health app, click on **Medical ID**, and click on **Edit**. Select **Edit medical ID** and click on **Add emergency contact.** Select a contact on your list, add your relationship with the contact and save the contact as one of your emergency contacts. You can add more than one emergency contact.

How to Show Previews on Lock Screen

You can see previews of your notifications on your lock screen if you follow these steps. Go to **Settings** on your phone, click on **Notifications**, click on **Show previews,** and select **Always**. This way, your lock screen will show you previews of messages, chats, mails, and other pending notifications.

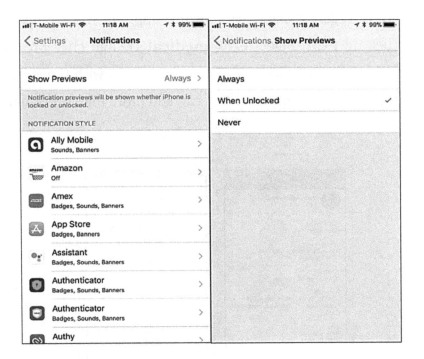

Customize Text Message Tones

Each iPhone comes with many text tones. You can make one to be your default text tone. Each time you get a text message, the default tone will sound.

Change the default text tone by navigating through the process below:

- Go to the **Settings** app.

- Tap **Sounds & Haptics**.

- Tap **Text Tone**.

- Swipe to browse the list of text tones (you can utilize ringtones as text tones; they're on this screen, as well). Tap a tone to hear it play.

- When you've discovered the text tone you need to utilize, tap it to put a checkmark on it. Your decision is automatically saved, and that tone is set as your default.

Set Your Notification Preferences

You can pick whether to show an app notification on the lock screen or if you'd only like it shown when your face has been recognized.

To customize this feature, go to **Settings > Notifications > Show Previews** to choose how content is or isn't shown on the lock screen; alternatively, go to **Settings > Notifications** to adjust the lock screen look.

How to Activate Siri

During setup, you have the option to set up "Hey, Siri!" voice activation. However, here is a guide on how to set "Hey, Siri" if not done the first time you powered on your phone.

Enable "Hey Siri"

- Open the Settings app on your phone.

- Click on **Siri & Search.**

- On the next screen, you get to choose either **Press Home for Siri** or **Listen for "Hey Siri."** You may toggle on the two options if you want to activate Siri using the two options.

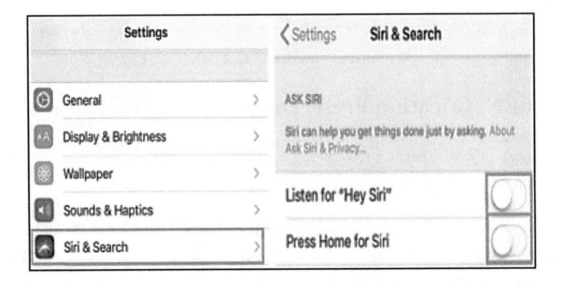

- Click on **Enable Siri** from the popup on your screen.

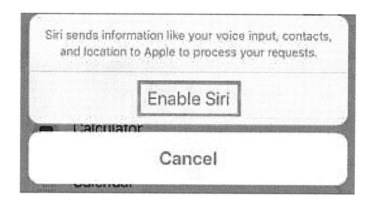

Train Siri to Recognize Your Voice

After enabling Siri, the virtual assistant will ask you to train it to identify your voice. Here is how to do so:

- Click on **Continue** on the Siri setup page.

- Say "Hey Siri" into your iPhone.

- Continue to follow the prompts on your phone. Repeat the prompts received on your phone to help Siri get familiar with your voice.

- Each time Siri gets the information it requested; you will see a checkmark on your screen.

- After repeating all the words shown on your screen, click on the **Done** button to begin using Siri.

Use "Hey Siri"

Here is how to use this virtual assistant.

- Make sure you are close to the iPhone.

- Say "Hey Siri" in a loud tone that the iPhone can pick up.

- Then tell Siri what you need at the time, like, "what is the weather like in Florida?" etc.

Change Siri's Language

Here is how to set your preferred language for Siri:

- Open the Settings app on your phone.

- Click on **Siri & Search.**

- Click on **Language**, then select the one of your choice.

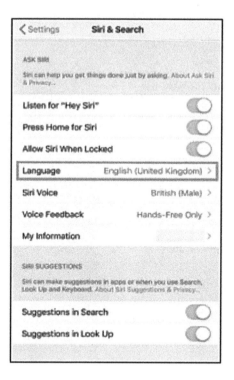

Secure Siri's Lock Screen

This option allows you to access Siri even when your phone is locked. With this option enabled, you can speak to Siri while your screen is locked. This can, however, reduce the security of your device.

- Open the Settings app on your phone.

- Click on **Siri & Search.**

- Toggle on **Allow Siri When Locked.**

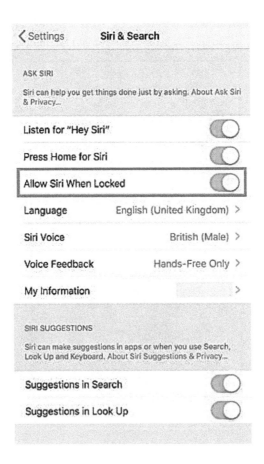

Change Siri Voice

You can set up Siri to speak in your preferred voice: male or female, American, British, etc. Here is how to:

- Open the Settings app on your phone.

- Click on **Siri & Search.**

- Click on **Siri Voice** and set up as desired.

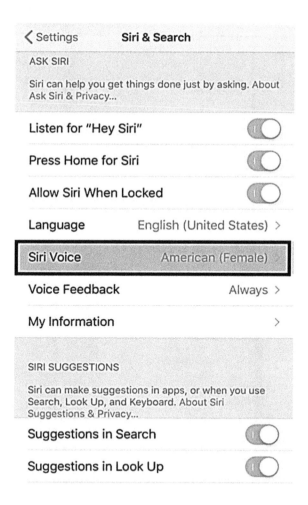

Change Language on iPhone

- On the Home screen, go to **Settings**.

- Next, tap **General**.

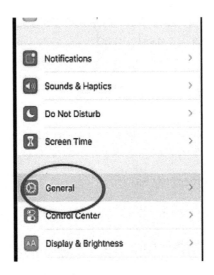

- Scroll down and select **Language & Region**.

- On the following screen, choose **iPhone Language**.

- Select your language from the rundown.

- A notification will require you to confirm the new language. Press the first option.

- After your iPhone updates the preferred language, it should automatically be showing the language you choose.

How to Use Memoji Feature

Create and Use Animoji or Memoji

The steps below will show you how to create an Animoji or Memoji.

- From the profile picture settings page, click on the **circle for photos** close to the name field.

- Then click on the ⊕ sign to design your own Memoji.

- After you must have created one, click on it to choose a pose for your Memoji and also use it as your profile picture.

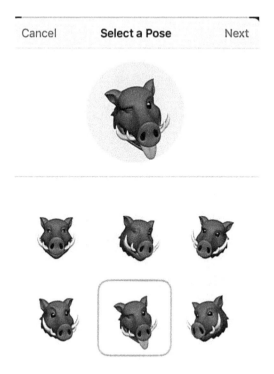

- If you prefer to use a non-personal image, you can select from available Animoji. The Animoji menu selection has several options, including sharks and mice.

- After you must have selected an Animoji or Memoji, scale it, then fit it into the circle.

- Choose a background color to complete your setup.

How to Use Siri Shortcuts

Add Siri Shortcuts

Siri now has an app that makes it easier and faster to assign functions to the virtual assistant.

- Click on the **Shortcut app** to launch it.

- After which, you click on **Create Shortcuts** to create a simple type of shortcut.

- Then click on **Create Shortcuts** to create a simple type of shortcut.

- With the **automation tab**, your device can intelligently react to context as they change. For instance, you can customize the shortcut to play a particular song each time you get home or design the button to automatically send your location to your partner once you leave the office at the end of the day.

- In the **Gallery** tab, you will find a range of predefined shortcuts to give you some inspiration in designing yours, or you can even make use of the predefined shortcuts.

How to Set Screen Time

This feature helps you to monitor how much time you spend on your device. On the screen time option, you will see the complete details on the hours spent on your iPhone.

Enable Screen Time

- Open the **Settings** app.

- Click on **Screen time.**

- Click on **Turn on Screen Time** to enable the feature.

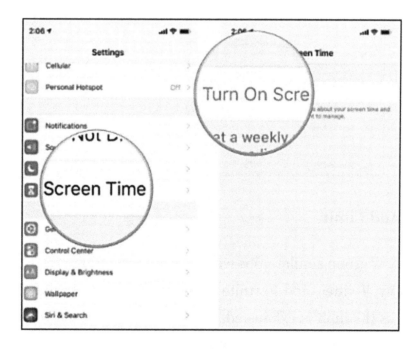

Set App Limits

Use the **App Limit** feature of the Screen Time to set the length of time you want to spend on certain apps. Follow the steps below to guide you.

- Go to the **Settings** app on your device.

- Then click on **Screen Time**.

- Navigate and click on **App Limits.**

- Then select **Add Limit.**

In iOS 13, you can now group similar apps with the same app limit. What this means is that you can bring Spotify, Twitter, and Fortnite together to have a combined total of 6 hours every day. As soon as the limit is exhausted, a splash will appear on your screen, notifying you that the set limit is exhausted. You will also get the option to ignore the deadline for the remainder of the day or 15 minutes only.

How to Create a New Reminder

The Reminders app allows you to create reminders, including attachments and subtasks. You can set alerts based on location and time. You can also be alerted when sending a message to someone. There are several things you can do with the Reminders app, which we will look at in this section.

Create a New List

Here is how to create a new list of reminders:

- Open the **Reminders** app.

- Click on **Add List** on the home page of the Reminders app.

- Input your preferred name for the list.

- Select color and an icon for your list to make it easy for you to tell the lists apart.

- Click on **Done**.

Edit a List

This is how to edit a list that you already created:

- Click on the list that you wish to edit.

- Click on the more button ···.

- Click on **Name and Appearance.**

- Then change the icon, color, or name of the list.

- Click on **Done.**

Create a Reminder

- Open the **Reminders app**.

- Click on the list that you wish to create a new reminder in.

- At the bottom left of your screen, click on **New Reminder.**

- Input your reminder in the text field.

- Click on **Done**.

Add an Attachment to a Reminder

After creating your reminder, here is how to add an attachment to that reminder:

- At the bottom left of your screen, click on **New Reminder.**

- Input your reminder in the text field.

- Now click on the camera icon ⬛ above the keyboard.

- Select your preference from the drop-down list: **Take a New Photo, Scan a Document,** or **Choose an Image From Your Phone's Photo Library**.

- Attach the image.

- Click on **Done** to save your reminder with the attachment.

Tag Someone in Your Reminder

- Open the Reminders app.

- Click on the list that you wish to create a new reminder in.

- At the bottom left of your screen, click on **New Reminder.**

- Input your reminder in the text field.

- Click on the ⓘ button to display the Details screen.

- Toggle on the option besides **Remind Me When Messaging**.

- With the toggle on, click on **Choose Person** and select from your contact list. You may see **Edit** if you already added someone and want to make changes.

- When next you are sending a message to the person, a reminder will pop up on your screen. It will also show it in the **Notification Center** and the **Lock Screen** until you complete it.

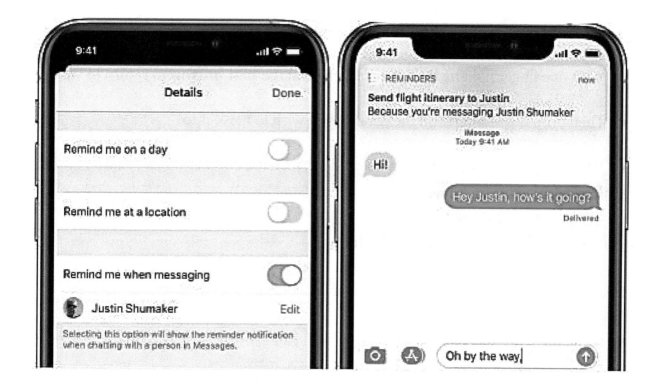

Move a Reminder to a Different List

- Open the **Reminders** app.

- Click on the list that has the reminder you want to move.

- Click on the reminder to open it.

- Then click on the **edit details button** ⓘ.

- Click on **List**, then select the new list you want to place the reminder in.

- Then click on **Done**.

Alternatively, to drag the reminder to a different list:

- Click and hold the reminder you wish to move with one finger.

- With one finger holding the reminder, click on the **List button** to return to your list.

- Then drop the reminder on the list that you wish to move it to.

- For multiple reminders, click on one, hold it, then use another finger to select the other reminders you wish to move.

Create a Scheduled Reminder

Here is how to create a scheduled reminder:

- Open the Reminders app.

- Click on the list that you wish to create a new reminder in.

- At the bottom left of your screen, click on **New Reminder.**

- Name your reminder.

- Then click on the ⓘ button by the name of the reminder.

- Toggle on the switch beside **Remind Me on a Day.**

- Click on **Alarm** and set your preferred date.

- If you want to be reminded at a specific time on the set date, toggle on the **Switch for Remind Me at a Time**, and set your preferred time.

- Click on **Done**.

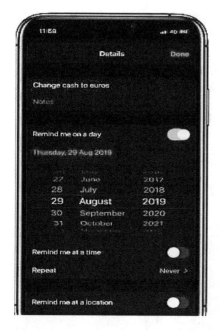 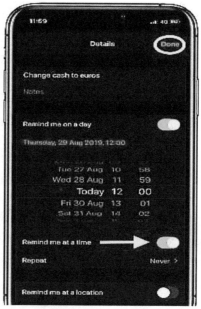

View Completed Reminders

After you mark a task as complete, the reminder will be removed from the Reminders app to make room for other active reminders. Follow the steps below to show completed reminders on the home page of the Reminders app:

- Open the Reminders app on your phone.

- Navigate to **My Lists** and click on a reminders list.

- Click on the **three-dot icon** at the top of your screen.

- Then click on **Show Completed** from the pop-up menu. To hide the completed reminders, come back here and click on **Hide Completed**.

 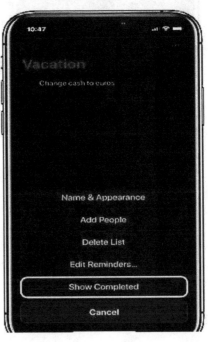

Delete a Reminder List

- Open the Reminders app on your phone.

- Swipe left on the list you wish to delete.

- Click on the **Delete** icon.

Group Different Lists Together

Your phone operating system allows you to group separate lists into one single group. This helps to make your reminder interface neat. It also makes it easy to organize related lists. For instance, if you have different lists containing dates like birthdates, anniversary dates, etc., you can group all of them into a **Memorable Date group**. Here is how to do this:

- From the Reminders app, click **Edit** at the top of the screen.

- Then click on **Add Group** at the bottom side of your screen.

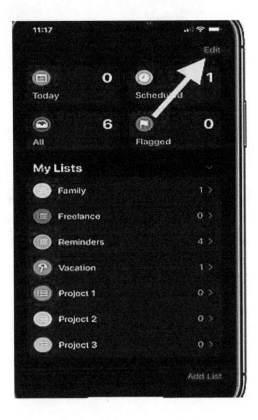 

- Title your group.

- Click on **Include**, then choose the reminder list that you want to include in the group by clicking on the green + button beside each list.

- Click on **New Group** to go back to the previous screen.

- Then click on **Create**.

- Click on a group to view the different lists inside.

- To delete a group, swipe left on a group, then click on the **delete** button. You will receive a prompt to confirm if you want to delete the group only or delete the group and all the lists in it.

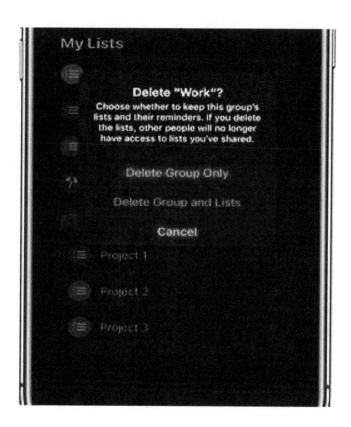

Share a Reminder

Here is how to share your list with someone:

- Open the Reminders app.

- Click on the desired list to open it.

- Then click on the **three-dot icon** ⬤ at the top of the screen.

- Click on **Add People** from the drop-down list.

- You will be prompted to input the email address of the receiver.

- Once done, click on **Add**.

- Then click on **Done**.

Change Reminder Priority

After creating a list and adding reminders to the list, you can set the priority of the reminder with the steps below:

- Open the desired reminder.

- Click the ⓘ icon at the top of the screen.

- Navigate down on the next screen and click on **Priority**.

- Choose your preference from the priority options presented: **None, Low, Medium, High.**

- Click on **Details** back arrow to return to the previous screen.

- Then click on **Done**.

How to Change Wallpaper

- Launch the **Settings** app on your iPhone.

- Go to **Wallpaper**.

- Select **Select wallpaper.**

- Select the photo or image you want to set as wallpaper.

- At the lower right of the phone, tap **Settings.**

- Tap **Set to locked screen / Set to home screen / Set for both**.

Move Home Screen Apps

Remove Applications From the Home Screen

If you, for example, on the home screen on an **App icon,** tap and hold it, select just the app widget (if available), in the last used documents appear to be. You can move the icon around the screen using the same procedure.

Main Menu

There are two different ways to remove an app from the Apple iPhone home screen.

Hold for one second **the app down** and then select from the drop-down menu the **reorder option app.** This means that all apps will start shaken on the screen, and in the corner of each one appears a symbol, traditionally a small **X**, allowing you to remove them individually.

The other way to access this screen is to hold down the app icon for at least two seconds: with each widget, the pop-up menu disappears, and the icon goes under your finger and can be removed.

Check Battery Level as a Percentage

How to Check Battery Percentage on iPhone SE (2020) in 4 Easy Ways

Method 1: Check Battery Percentage in Status Bar

Step 1: Launch the Settings app, scroll down, and tap on **Battery** or alternatively, long press on the Settings app, and then tap **Battery.**

<u>Step 2</u>: Turn on the **Battery** Percentage **toggle.**

That's it. You should now see your iPhone SE's battery percentage right in the status bar, as you can see in the screenshot below.

Method 2: Check Battery Percentage in Control Center

This is probably the next most convenient method.

Step 1: Swipe down on the upper-right corner of the screen in portrait or landscape mode to access **Control Center**.

Step 2: You should now see the hidden battery percentage along with other hidden Status bar icons for Bluetooth and Location Services.

That's it. You can Swipe up from the bottom of the screen or tap the screen to close the Control Center.

Method 3: Check Battery Percentage By Asking Siri

You can also ask Siri to tell you the charge remaining on your iPhone SE.

Step 1: Press and hold on the side button or if you have the Listen for "Hey Siri" enabled in the Settings.

Step 2: Then just say: "Hey Siri, please tell me the current charge on my iPhone." It will promptly tell you the battery percentage.

Method 4: Check Battery Percentage in Today View

You can also view your iPhone SE's battery percentage in the Today view by enabling the Batteries widget.

Step 1: Swipe down from the top of the screen on the first **Home screen page** or the **Lock screen** to access the **Notification Center**; swipe right to access the **Today View**.

Step 2: Scroll to the bottom and tap **Edit**, then hit the **plus** (+) sign next to **Batteries** under **More Widgets** section to add it. Tap **Done** in the top right corner.

You should now be able to view the battery percentage indicator in the **Batteries** widget in the Today View, as you can see in the screenshot below.

Calling, Video Calling, Taking Pictures, Sending Messages, and Adding Numbers

Calling

How to Call a Phone Number

The major essence of having a phone is to be able to make calls.

To make a call:

- Tap on the phone icon on your home screen.

- Tap on the **Keypad** shortcut.

- Enter the required phone number and tap the call icon.

How to Turn Call Waiting On/Off

The call waiting function, when turned on, allows you to answer a new call without ending an ongoing call.

This function can be turned On/Off by:

- Tap on **Settings** on the Home screen of your device.

- On the settings page, scroll down and tap on **Phone.**

- Tap on **Call Waiting** on the Phone page.

- Toggle the indicator beside **Call Waiting to turn it On/Off.**

- Tap on the **home button** to return to the home screen.

How to Save a Voicemail Number

You can save a voicemail number, which will make it easier for you to call and listen to your voice messages.

To save a voicemail number:

Once you insert your SIM into your phone, the voice mail number is saved automatically.

How to Cancel All Call Divert

If at any point you no longer wish to divert calls coming to your phone, you need to cancel all call divert.

To cancel all diverts:

- Tap on the **Phone icon** on the home screen of your device.

- Tap on the **keypad icon.**

- Type **##002#** and tap on the call icon.

- A display with all the divert options will appear on your screen; tap on **Dismiss.**

- Tap on the **home button** to return to the home screen.

How to Turn On/Off Call Announcement

You can set your iPhone to announce the contact that's calling you when a call comes through your device. If the number is not saved on your address book, your device will call out the digits.

To set call announcement:

- Tap on **Settings** on the Home screen of your device.

- On the settings page, scroll down and tap on **Phone.**

- Tap on **Announce Calls** on the Phone page.

- On the next page, select **Always.**

- If you don't want your device to announce callers ID, select **Never.**

- Select **Headphones & Car** to turn on the function when your iPhone device is connected to a headset or your car.

- Tap on **Headphones Only** to turn on the function when your device is connected to a headset.

- Tap on the **home button** to return to the home screen.

Taking Pictures

How to Setup the Camera?

Your camera is already working once you start using your iPhone SE 2020. However, you may want to make certain changes to suit yourself. If you want to avoid having to set your camera to taste every time, you can preserve your camera settings. To do so, go to your phone settings and click on **Camera**. Under the camera settings, select **Preserve settings**, and you can choose to turn on the camera settings you want to preserve. Under the camera menu in your phone settings, you can turn on **Grid** to help you straighten your shots when you capture a picture.

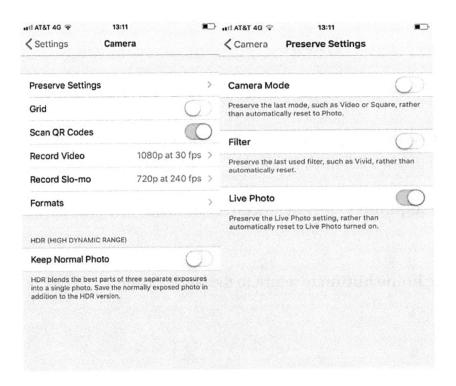

How to Use the Front-Facing Camera?

Selfies are a thing these days. It is awkward using your main camera instead of your front-facing camera for a selfie. To use your front-facing camera, open your phone camera and click on the camera icon with two circular arrows around it at the bottom of your screen. You will observe the change in camera modes as your face will show on your screen as it appears in the camera. You can make the adjustments you want and take the snapshot. You can also edit your selfie using the apps on the phone.

How to Take a Live Photo?

Live photos are similar to GIFs. They are best for capturing moments that cannot be best explained in a still shot. To capture a live photo, ensure the **Live photos** icon at the top of the screen is set to on. It is off if the icon has a line through it. In live photo mode, your phone will capture 1.5 seconds of what happens before and after you click the shutter button. You should make sure you're set before clicking the shutter button.

 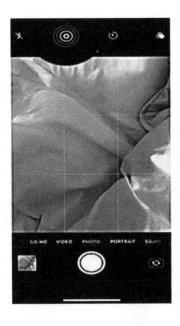 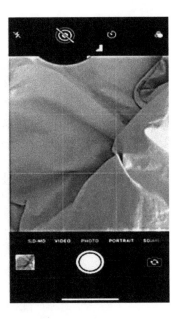

How to Take a Panoramic Photo?

Panorama helps you capture a wide series of pictures as a single picture. To take panoramic pictures, open your phone camera and swipe left till panorama appears at the bottom of your screen. Position your phone at the start point and click on the shutter button. Move the phone from left to right till the process is complete. You can also click on "stop" to manually stop it if you have captured what you want.

How to Adjust the Exposure?

The exposure of the picture determines how much light the subject of the picture will receive. It helps emphasize certain aspects of the picture. To adjust the exposure of the picture, click on the subject on the picture and swipe up or down to adjust the exposure. Swipe up to brighten the picture and swipe down to darken the picture.

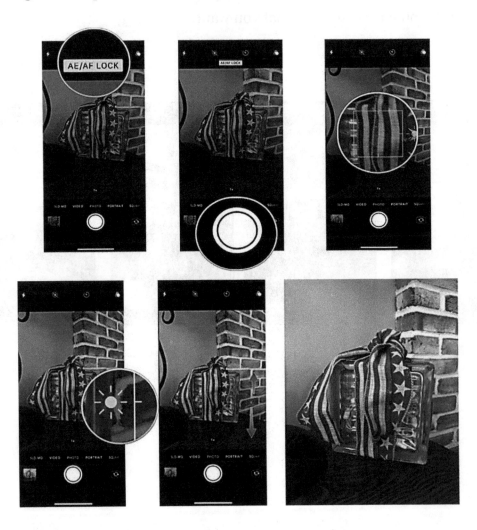

How to Capture With Time-Lapse?

Time-lapse capture helps you take a hands-free picture. It comes in when you want to appear in a regular picture rather than a selfie. To capture with time-lapse, go to your camera setting, locate the time-lapse, and set your preferred time-lapse (3 seconds or 10 seconds). After setting your time-lapse, you can click on the shutter button and position yourself in the area of view of your camera. Your phone will take burst photos of about 10 pictures, and you will select the one you prefer. To get the best of this mode, place your phone on a still surface like a tripod stand.

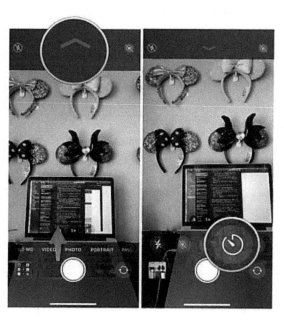

Adding/Deleting Contacts

Add Contacts

- Open the **Contacts** app.

- Then click on $+$ > **Add** icon at the top right side of your screen.

- Enter the details of your contact, including the name, phone number, address, etc.

- Once done inputting the details, tap **Done** to save.

Merge Similar Contacts

- Open the **Contacts** app on your home screen.

- Click on the contact you want to merge and click on **Edit.**

- At the bottom of the screen, click on **Link Contact.**

- Choose the other contact you want to link.

- Click on **Link** at the top right side of the screen.

Copy Contact from Social Media and Email Accounts

- Open the Settings app and click on **Accounts and Password.**

- Click on **the account**, e.g., Gmail.

- Toggle on **Contacts.**

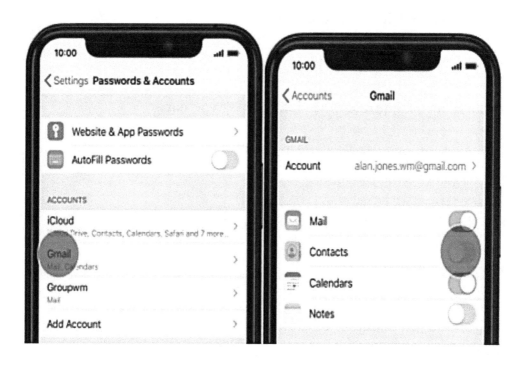

Add a Caller to Your Contact

- On your call log, click on a phone number.

- You will see options to **Message, Call, Create New Contact,** or **Add to Existing Contact**.

- Select **Create New Contact.**

- Enter the caller's name and other information you have.

- At the top right hand of the screen, click on **Done**.

Add a Contact After Dialing the Number on the Keypad

- Manually type in the numbers on the phone app using the number keys.

- Click on ⊕ on the left side of the number.

- Click on **Create New Contact.**

- Enter the caller's name and other information you have.

- Or click on **Add to Existing Contact**. Find the contact name you want to add the contact to and click on the name.

- At the top right hand of the screen, click on **Done**.

Import Contacts

The iPhone allows you to import or move your contacts from your phone to the SIM card or SD card for either safekeeping or backup. See the steps below:

- From the Home screen, click on **Settings.**

- Select **Contacts.**

- Click on **Import SIM Contacts.**

- Chose the account where the contacts should be stored.

- Allow the phone to import the contacts to your preferred account or device.

Delete Contacts

Follow the steps below to delete a contact:

- From the Home screen, tap on **Phone** to access the phone app.

- Select **Contacts.**

- Click on the contact you want to remove.

- You will see some options, select **Edit.**

- Move down to the bottom of your screen and click on **Delete Contact.**

- You will see a popup next to confirm your action. Click on **Delete Contact** again.

- The deleted contact will disappear from the available Contacts.

Sending Messages

Set up your Device for iMessaging

- From the Settings app, go to **Messages**.

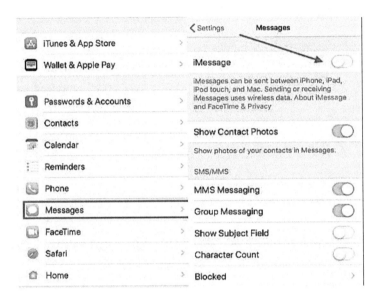

- Enable **iMessages** by moving the slider to the right.

Set up Your Device for MMS

- From **Settings**, go to **Messages**.

- Enable **MMS Messaging** by moving the slider to the right.

Compose and Send iMessage

- From the Message icon, click on the **New Message** option at the top right of the screen.

- Under the "**To**" field, type in the first few letters of the receiver's name.

- Select the receiver from the drop-down.

- You will see iMessage in the composition box only if the receiver can receive iMessage.

- Click on the **Text Input Field** and type in your message.

- Click on the send button beside the composed message.

- You will be able to send video clips, pictures, audios, and other effects in your iMessage.

Compose and Send SMS

- From the Message icon, click on the new message option at the top right of the screen.

- Under the "**To**" field, type in the first few letters of the receiver's name.

- Select the receiver from the drop-down.

- Click on the "**Text Input Field**" and type in your message.

- Click on the send button beside the composed message.

Compose and Send SMS With Pictures

- From the Message icon, click on the **New Message** option at the top right of the screen.

- Under the "**To**" field, type in the first few letters of the receiver's name.

- Select the receiver from the drop-down.

- Click on the **Text Input Field** and type in your message.

- Click the **Camera** icon on the left side of the composed message.

- From **Photos**, go to the right folder.

- Select the picture you want to send.

- Click **Choose** and then **Send.**

Create New Contacts From Messages on iPhone

- Go to the Messages app.

- Click on the conversation with the sender whose contact you want to add.

- Click on the sender's phone number at the top of the screen, then click on **Info**.

- On the next screen, click on **the arrow** by the top right side of your screen.

- Then click **Create New Contact**.

- Input their name and other details you have on them.

- At the top right hand of the screen, click on **Done**.

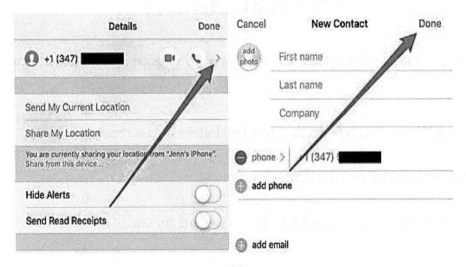

Hide Alerts in Message app on your iPhone

- Go to the **Message app** on your iPhone.

- Open the conversation you wish to hide the alert.

- Click on ⓘ at the upper right corner of the page.

- Among the options, one of them is **Hide alerts**; move the switch to the right to turn on the option (the switch becomes green).

- Select **Done** at the right upper corner of your screen.

Set a Profile Picture and Name in iMessages

Setting a profile picture and name in iMessages saves users from first saving your contact details to decipher who sent a message. Follow the steps below to set this up.

- Open the messages app.

- Click on the ⬤ icon at the upper right corner of your screen.

- Then click on **Edit Name and Photo.**

- Select a profile picture and type in your last and first names.

- You can either create your own Memoji to use as your profile picture or select from available Animoji.

Specify View for Your Profile Picture and Name in iMessages

You can choose who you will like to be able to view your profile picture and name. This setting is used to limit users who can access your details.

- Open the messages app.

- Click on the ⬤⬤⬤ icon at the upper right corner of your screen.

- Then click on **Edit Name and Photo.**

- Enter your name and choose your profile picture.

- Turn on the switch for **Share Name and Photo,** then select from the three options available for sharing your name and profile picture:

  ➢ Use the **Contacts Only** option to share details with only persons whose numbers you have saved on your smartphone.

  ➢ **Anyone** option gives access to everyone that has your contact details.

  ➢ If you want the system always to prompt you to choose who to share with, then click on the **Always Ask** option. Whenever you open a new message, you will see a pop up on your screen asking for permission to share details with the sender. To share your details, click on **Share;** if otherwise, click on **X** to refuse and shut down the message.

Share Name and Photo ⬤

SHARE AUTOMATICALLY

Contacts Only

Always Ask ✓

Anyone

You will be prompted before updated name and photo are shared.

Video Calling

Face Time

There is a built-in FaceTime app on your iPhone SE. FaceTime is also integrated into the iPhone SE's phone app. This app allows you to not only hear but see the people you are talking to. Use the front-facing camera of your phone to talk face to face, or switch to the rear camera so that the people on the call can see what is in front of you. All you need is an active Wi-fi connection or cellular data. Also, your call recipient has to have an iOS device that is FaceTime activated.

Set Up FaceTime

- Open the Settings app on your device.

- Click on **FaceTime.**

- Move the switch beside **FaceTime** to the right.

- Click on **Use Your Apple ID for FaceTime.**

- Then click on **Sign In.**

- Or, click on **Use Other Apple ID**, then enter the credentials you wish to use.

- Choose the phone number that you want people to be able to reach you on when using FaceTime. You can also register an email address to use for FaceTime.

- Tap the email address or phone number that you wish to use as your **caller ID.**

- Press the Back button to return to the previous screen or tap the Home button to go home.

Turn on FaceTime

Here is how to turn on FaceTime on your phone.

- Launch the FaceTime app.

- Sign in using your Apple ID.

- FaceTime will automatically register the phone number(s) linked to your phone.

Disable FaceTime

While FaceTime could be fun to use, you may not want to be interrupted or surprised by a prompt to accept an audio or video call. So, here is how to turn off FaceTime.

- Open the Settings app on your device.

- Click on **FaceTime.**

- Move the switch beside **FaceTime** to the left to turn it off. The gray color shows that the feature has been disabled.

- To enable it in the future, return to step 3 and move the switch to the right.

Change Your FaceTime Email Address

If you decide to use a different email address other than the one currently registered on the FaceTime app, follow the steps below to change to the new email address:

- Open the Settings app on your device.

- Click on **FaceTime.**

- If you are not already signed into FaceTime using your Apple ID, Click on **Use Your Apple ID for FaceTime.**

- Then click on **Sign In.** Or, click on **Use Other Apple ID,** then enter the credentials you wish to use.

- Tick the boxes to the left side of the email address(es) you want to remove from FaceTime on this phone. The removed email will have no checkmark.

- Then click on the email address(es) you want to use for FaceTime so that a checkmark appears by the side.

Note: you can register different emails on different devices. For example, you can have a yahoo mail address on your iPad and have a Gmail address on your iPhone SE. You can also add more than one email address so that people can reach you on any of the registered email addresses.

Make a FaceTime Video Call on iPhone

- Open the **FaceTime** app on your phone.

- Click on the ⊕ button.

- Enter the phone number, email address, or the name of the person you want to call. As you type, matching contacts will show up on your screen. Click on the person you want to call. Or click the ⊕ button to choose someone from your Contacts.

- Enter more numbers, names, or email addresses if you are going to make a group call.

- Click on **Video** to begin the call.

Make a FaceTime Audio Call on iPhone

- Open the FaceTime app on your phone.

- Click on the ⊕ button.

- Enter the phone number, email address, or name of the person you want to call. As you type, matching contacts will show up on your screen. Click on the person you want to call. Or click the ⊕ button to choose someone from your contacts.

- Enter more numbers, names, or email addresses if you're going to make a group call.

- Click on **Audio** to begin the call.

Make FaceTime Call with Siri

You can also use Apple's virtual assistant to make a FaceTime call.

- Say **"Hey Siri"** to activate the virtual assistant. Alternatively, press and hold the home button until the Siri prompt appears.

- Then say "FaceTime..." followed by the name of the person you want to call. Alternatively, say "FaceTime..." and wait for Siri to ask for the name of the recipient.

- Then wait for your call to connect.

Switch From FaceTime Video to FaceTime Audio

You can always switch between video call and audio call in the FaceTime app at any time during a call.

- Answer or start a FaceTime call in the usual way.

- Click on your screen to display the call controls.

- Swipe up on the control panels.

- Then click on **Camera Off.**

- Click on the Camera settings to return to Video call.

Here is another way to switch to an audio call on your FaceTime app for iPhone SE:

- Answer or start a FaceTime call in the usual way.

- Press the **Home** button to exit the FaceTime app. The person on the other end will continue to hear you but will not be able to see you.

Switch From Regular Call to FaceTime

While on a regular call with someone who is also using an iPhone, you can switch to FaceTime video chat without hanging up. Here is how to do this:

- Open the Phone app and put a call through to someone.

- While on the call, look at your phone; you will see the call menu that appears when making a call.

- Click on the **FaceTime** icon to switch to FaceTime video call.

Mirror a FaceTime Call to an AirPlay 2 Compatible Smart TV or an Apple TV

You can use the AirPlay feature on your smartphone to mirror a FaceTime call on a compatible smart TV. Before you start, ensure that your iPhone and the smart TV are connected to the same Wi-fi network.

- Swipe up on the home screen of your phone to open the Control Center.

- Click on **Screen Mirroring.**

- Choose your Apple TV or other AirPlay 2 compatible smart TV from the displayed list.

- An AirPlay passcode may pop up on your TV screen. Input the passcode on your smartphone.

- After the screen mirroring connection is established, open the FaceTime app, and start your video call. The video will now show on your mirrored device.

Stop Mirroring FaceTime Call on Your Apple TV

Here is how to stop mirroring your iPhone on your compatible smart TV.

- Swipe up on the home screen of your phone to open the Control Center.

- Click on **Screen Mirroring.**

- Then click on **Stop Mirroring.**

- Or, if mirroring to an Apple TV, simply click the Menu button of your TV remote to stop mirroring.

Make FaceTime Call From Your Favorites

If you frequently call the same people on your contacts list, you may save time by adding these people to your Favorites list on your phone app. You can then click on the person's name to initiate a FaceTime call right away. Follow the steps below to add a contact to your Favorites as a FaceTime contact.

- Open the Phone app on your smartphone.

- Click on the **Favorites** tab at the bottom end of your screen.

- Click the ⊕ icon at the left top side of your screen.

- Search for the contact or choose the contact from the displayed list.

- Click on **Video** in the menu on your screen.

- Tap **FaceTime** from the displayed options.

How to Use Social Network

How to Install Facebook on Your iPhone Device

To use Facebook on your device, you need to have it installed. To install Facebook, you need to connect your device to the internet and have your Apple ID.

To install Facebook:

- Tap on the **App Store app** on the home screen of your device.

- On the app store page, tap on the **search icon** at the bottom right side of your device.

- Type **Facebook** on the search bar and click on the first option.

- On the Facebook page, tap on **GET** and follow the instructions on the screen of your device to get it installed.

- Tap on the **home button** to return to the home screen.

Follow the above steps to install every other relevant Apps you want.

How to Download Apps

Do you think of where you can get those stunning and beautiful applications that give so much fun? It is the App Store. In the app store, you get to purchase lots of applications and as much as you want. While some are free, you might be paying for others. Note that you are not limited to applications alone in the app store; you can as well access games, books, and more.

How to Find an App

To find an application, use the search tab at the lower part of your screen. You can as well search through categories like games, books, applications, and more if you do not know what exactly you are looking for. To ease some stress, you can make use of the Siri voice control to do your search. Just hold the Home button till Siri beeps.

How to Buy, Redeem, and Download an App

When you click on an application, you will be asked to either download for free or make a payment. If the application is paid, know that you will be purchasing with your payment details in your Apple ID. However, if the app displays what looks like a cloud, this means you have previously installed the app. Hence, you can install it again for free.

App Store Settings

You can set up your app store with different options by going to **Settings** and then to **iTunes & App Store**. This allows you to:

- View and edit your account.

- Change your Apple ID password.

- Sign in with a new or different Apple ID.

- Subscribe and turn on iTunes Match.

- Turn on automatic downloads for books, music, tv, shows, movies, and more.

How to Block a Telephone Number

How to Block a Phone Number?

If you don't want to receive voice calls or messages from number(s), you can block them. When a blocked number calls you, your number will give them a busy signal.

To Block a Number

- Tap on the **Phone** icon on the home screen of your device.

- Tap on **Recent**.

- Tap on the **information icon** next to the number you wish to block.

- Scroll down and tap on **Block This Caller**. A prompt will appear notifying you that you will not receive calls and messages from the number when blocked.

- Confirm by tapping on **Block Contact.**

- Tap on the **home button** to return to the home screen.

How to Customize Voiceover

VoiceOver is a function that reads out the screen so that even people with visual impairments can use the iPhone. The selected screen display will be read by the iPhone on behalf of the user.

You can also change the speed and pitch at which the screen display is read out using VoiceOver. Detailed settings can be changed in **Settings app > Accessibility > VoiceOver**.

Set Up Emergency Medical ID

- Open the **Health** app.

- Touch **Medical ID** at the bottom-right corner.

- Next, tap on **Create Medical ID** to start adding your health info.

- On the following screen, enter all your medical information, including allergies, well-being conditions, emergency contact details, and any helpful notes. This will be valuable if there's an occurrence of an emergency, and anyone around you can rapidly access this information.

- After you are done with adding the details, switch ON **Show When Locked**. This feature is optional yet exceptionally recommended. The reason is that all the info you have entered will be noticeable to others regardless of whether your iPhone is encrypted.

That is all; you would now be able to leave the Health app and lock your iPhone. You can confirm whether it is working or not by swiping up to the Passcode screen and afterward tapping on **Emergency** and then tap on **Medical ID**.

Battery Tips

The iPhone battery tends to run down fast because of the different functions that the iPhone offer. I have compiled a list of things you can do to save your battery life. You do not need to perform all the tips stated here; pick a few that you are okay with.

Set Optimized Battery Charging

Optimized Battery Charging helps to reduce wear on your phone battery and also improve the battery lifespan by reducing the time your smartphone spends fully charged. With this feature enabled, your device will stop charging past 80% in most cases. The processor is designed to activate this feature only when it predicts that the phone will be connected to a charger for a long time.

- From the Settings app on your smartphone, click on **Battery.**

- Click on **Battery Health.**

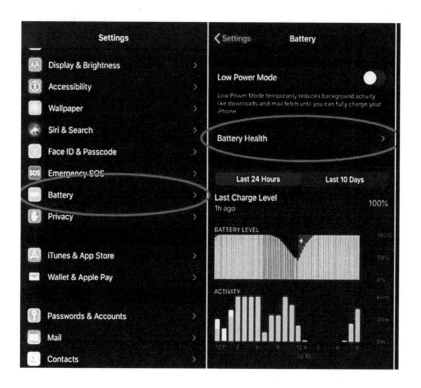

- On the next screen, you will see data showing the maximum battery capacity for your phone battery, an indication of the level of degradation, and the option to enable **Optimized Battery Charging**.

- Navigate to **Optimized Battery Charging** and move the switch to the right to enable this feature.

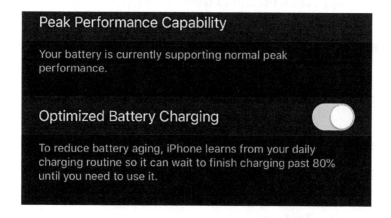

Stop Background Apps Refresh

One feature built to make your iPhone not only smart but also ready for you to use when needed is the Background App Refresh. The work of this feature is to look out for apps that you use often and the period in the day that you use these apps and then carry out automatic updates on the app so that you can have the latest information the next time you launch these apps. Follow the steps below to disable this feature on your smartphone.

- Go to the settings app and click on **General.**

- Then click on **Background App Refresh**.

- Move the slider to the left to disable this feature for the whole apps or disable the feature for select apps.

Disable Auto Update of Apps

You may have set up the option to automatically update your apps as soon as there is a new version, and this can drain the battery life. Follow the steps below to disable this feature and manually update your apps.

- From the settings app, click on the **iTunes & App Store**.

- Click on the button beside **App Updates** to disable the auto app update feature.

Extend the Device Battery Life

Some apps and services on the iPhone draw lots of power, which will drain the battery life faster. You can turn on low power mode to reduce power consumption.

- From **Settings**, go to **Battery**.

- Move the switch beside **Low Power Mode** to the right to enable.

Disable Auto App Suggestions

This feature makes use of your location service to discover your area and suggest apps that you may need based on your location. While it is a cool feature, it can, however, drain the battery. The steps below will show you disable the feature.

- Open the Settings app and click on **Siri & Search.**

- Under **Siri Suggestion,** toggle off **Suggestions in Search** and **Suggestions in Look Up.**

- You may also toggle off **Suggestions** on the **Lock Screen.**

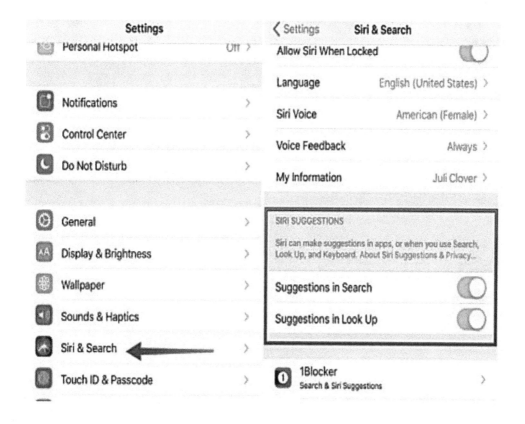

Enable Auto-Brightness

The auto-brightness feature makes your phone to adjust its brightness based on the lighting condition per time. The feature will save you battery life as it controls the energy used on the phone.

- From the Settings app, click on **Accessibility.**

- Then click on **Display & Text Size**.

- Scroll down and toggle off **Auto-Brightness**.

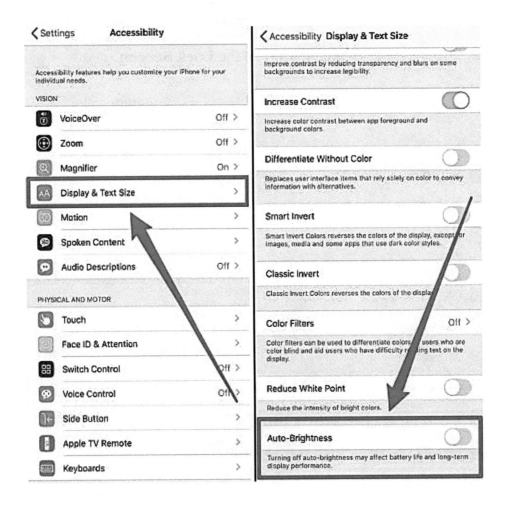

Reduce Screen Brightness

The brighter your screen is, the more power it consumes. You can control your phone's brightness using the slider on your iPhone. When you need to save battery, ensure that the brightness of the screen is at its lowest.

- From the settings app, click on **Display & Brightness.**

- Then use the slider to reduce the brightness by pulling to the left.

Stop Motion and Animations

Although this feature is cool, however, disabling it will help to save battery life.

- From the settings app, click on **General.**

- Then click on **Accessibility.**

- Select **Reduce Motion**.

- Move the slider beside **Reduce Motion** to enable it.

Disable Wi-Fi When Not in Use

When not using the Wi-Fi connection, it is advisable to disable it so it does not drain your battery.

- From the settings app, click on **Wi-Fi**.

- Then move the slider beside Wi-Fi to disable it.

- You can also disable it through the control center. Go to the control center and click on the **Wi-Fi** button until it turns grey.

Locate the Battery Draining Apps

You can find apps using a significant percentage of your battery through a feature called **Battery Usage**. Follow the steps below to use this step.

- From the Settings app, click on **Battery**

- The apps will be displayed according to the battery usage.

- At times, there may be notes under each app to show reasons the app consumed so much battery and fix it.

Ensure that Personal Hotspot is Disabled

When the hotspot is on, your iPhone becomes a hotspot that shares its cellular data with other devices in range. While it can be useful, it can also drain your battery. Whenever you are done with the hotspot, remember to disable it.

- Go to the Settings app.

- Click on **Personal Hotspot**

- Disable the option using the button beside it.

Disable Bluetooth

Bluetooth is another function that transmits data wirelessly and drains battery while doing so. To save battery life, put on Bluetooth only when needed. To either disable or enable your device Bluetooth, go to settings, and click on **Bluetooth.**

Disable Location Services

The iPhone comes with a built-in GPS, which is quite helpful for locating nearby restaurants, stores, etc., as well as finding directions. This app needs to send data over a network that usually tells on the battery of your device. Whenever you are not making use of the location services, you can disable it with the steps below:

- Go to the Settings app.

- Click on **Privacy.**

- Then select **Location Services**.

- Toggle off **Location Services** and then click on **Turn Off** to disable the option.

- Alternatively, you can navigate down the screen to select apps that should not have access to location services.

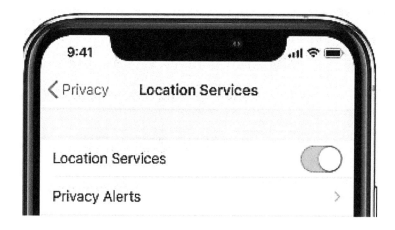

Disable Cellular Data

Similar to Bluetooth, when using 4G, 5G, LTE, and other cellular connections that have fast transfer speeds, they tend to drain your phone battery. They even consume more power when you are using it heavily, like when making HD calls or streaming videos. I know that cellular data is essential, which is why you should disable it only when you need to save battery life.

- From the settings app, click on **Cellular.**

- Go to **Cellular Data** and move the switch to the left to disable.

Note: Turning off cellular data will not affect your Wi-Fi connection.

Disable Data Push

You may have configured the email settings to automatically download messages to your smartphone as soon as they get to the email server. It is crucial to be current on your email folder; however, when you continuously download like this, it can drain your battery faster. Rather than the automatic update, you can go to the Mail app and manually refresh the app to receive new messages. The steps below will show you disable the data push feature:

- From the settings app, click on **Passwords & Accounts.**

- Or go to **Mails** from the settings app and click on **Accounts.**

- Then click on **Fetch New Data**.

- Go to **Push** and move the switch to the left to disable.

Set Emails to Download on Schedule

If you do not want to refresh your email manually, you can schedule the emails to download at a specified time. This is a balance between the two steps above; while you will not have to refresh your mail app manually, you will also not get an instant update. This method will still help to achieve the end goal, which is to save battery life.

- From the Settings app, click on **Passwords & Accounts.**

- Then click on **Fetch New Data**.

- Navigate to the bottom and choose your options. The longer the time between checks, the longer your battery life is preserved.

Set up the Screen to Auto-Lock Sooner

Auto-locking your screen helps to save your battery life. As long as your phone has something it is displaying, it will be taking out of the battery life. While I will advise that you select any of the options that suit you, do not choose **Never,** as that will drain your battery life.

- From the settings app, click on **Display & Brightness**.

- Then click on **Auto-Lock**.

- Select any of the options from 30 seconds to 5 minutes.

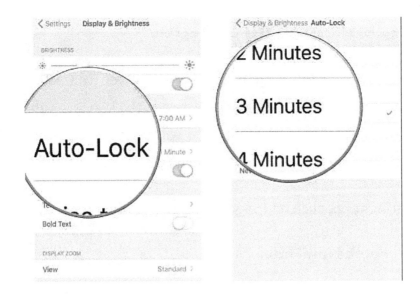

Disable Fitness Tracking

The fitness tracking feature on the iPhone is used to track your steps as well as other fitness activities. It is very beneficial, especially when you are trying to get into shape, but this app also drains the battery. You can disable the feature whenever you are not using it.

- From the settings app, click on **Privacy.**

- Then select **Motion & Fitness**.

- Move the switch beside **Fitness Tracking** to the left to disable this feature.

Disable AirDrop When Not in Use

AirDrop is the wireless file sharing feature of the Apple devices. To use Airdrop, you need to enable both Bluetooth and Wi-Fi and prepare your phone to locate other airdrop enabled devices. This makes use of more battery, and it is advisable to disable when not in use.

- Go to the control center, then click on **AirDrop.**

- Click on **Receiving Off** to disable the feature.

- You can also go to the settings app, click on **General,** select **AirDrop** and change **Receiving Off** or **Contacts Only.**

Disable Automatic Upload of Photos to iCloud

The photo app is set to upload your photos to your device's iCloud account automatically. Disable the auto-uploads and attempt to upload only when you have a full battery or when moving from your computer. Follow the steps below to check if your photos are always uploaded to iCloud.

- Go to the settings app.

- Click on **Photos.**

- Then click on **iCloud Photos**.

Stop Sending Diagnostic Data to Developers or Apple

The diagnostic data tell the developer how your smartphone is performing to help them produce better products. You have the option to enable this feature when setting up your device. Whenever you need to save battery life, follow the steps below to disable this option.

- Go to the Settings app.

- Click on **Privacy.**

- Then go to **Analytics** and shift the sliders to the left to disable this feature.

Disable Vibrations

When you place your device on vibration, the phone vibrates at every notification that comes into the device. The whole process involved in this causes the battery life to go low. Follow the steps below to turn off vibrations:

- Go to the Settings app.

- Click on **Sounds & Haptics**.

- Then move the slider beside **Vibrate on Ring** to the left to disable.

Other Helpful Tips to Improve the Longevity of Your iPhone Battery

Several factors can reduce the life of your battery, like leaving your phone plugged in even after the battery is fully charged. I have compiled the list below to help you prolong your battery life.

- Do not wait for your battery to drain before charging it. Do not let the phone battery drop to below 20% before you charge.

- Do not expose the smartphone to excess heat. Avoid charging your device in a scorching environment.

- If, for any reason, you do not intend to use your phone for about a week and above, ensure that the battery goes below 80% but not below 30 percent. Then shut down the phone properly before setting aside.

- Moving your phone quickly from a very hot to a very cold condition can affect the health of the battery.

- You do not need to always fully charge your phone, as it can aid in damaging your battery.

Troubleshooting Common Problems on Your iPhone SE

iPhone Won't Turn ON

Whenever you try to turn ON your iPhone, and it doesn't power On, it is either a software or hardware problem.

Whenever you encounter this issue, the first thing you should do is troubleshoot your device's software. Your iPhone may not power on if the software has crashed.

Solution

The fastest way to troubleshoot your software is to force reset your iPhone SE. You do this by pressing and quickly releasing the Volume Up button. Next, press and immediately release the Volume down button, then press and hold the Power button until the device restarts.

Another way to do this is to press and hold down both the Power and Home button together until the iPhone SE turns off and reboots. This action shuts down the iPhone completely.

If the problem stops after carrying out a forced reset, go ahead and continue using your device. However, if the problem keeps recurring, you will need to reset All Settings. A non-invasive software fix is excellent because, as it resets everything in your Settings App, it does not affect your personal information. Follow these steps to achieve this:

- First, tap **Erase All Content and Settings**. Next, enter your Passcode, and tap **Erase iPhone** to confirm. Your iPhone will restart once the reset has occurred. It fixes any glitch you may be having with your software.

- If you still keep encountering the same problem after following the steps above, what you should do as a matter of last resort is to carry out a DFU Restore. The DFU can be carried following the steps outlined below:

  ➢ First, connect your iPhone SE to a PC with iTunes installed.

  ➢ Next, press and hold the **Power and Home** button at the same time for about 10 seconds.

  ➢ After 10 seconds, still keep holding the **Home button** while you release the Power button.

  ➢ You should see detailed information on iTunes on your PC about your device being in recovery mode.

What to Do When Your iPhone SE Screen is Frozen

Several factors can freeze the screen of your device. The frozen screen can easily be solved by carrying out a hard reset. This can be achieved by holding the Power and Home buttons together until the Apple logo appears and disappears or until the screen turns black.

iPhone SE Won't Charge

If your iPad refuses to charge, it is usually as a result of a software or hardware problem. Your approach to solving this problem should be to first troubleshoot the software before anything else is done.

To Troubleshoot the Software, Hard Reset

Your iPhone by holding both the Power and Home buttons together for 10 seconds. You will notice the Apple logo disappears. A software crash is the most common cause of your battery's failure to charge. If you still keep experiencing the same problem, check the cable frames for discoloration. Use another lightning cable to charge it and check if there are any difference.

If it still doesn't work, clean the lightning port with a clean toothbrush dipped in methylated spirit.

iPhone Screen Won't Rotate

This is usually a software problem. Most times, it is often because your Orientation lock is turned on. There are different ways to turn this off.

Turning It Off From the Control Center

- First, open the Control Center.

- Find the orientation lock icon.

- Tap **the icon** to turn off your device's orientation lock.

If, after turning off your device's orientation lock, your iPhone still does not rotate, it means the device orientation app has crashed and should be closed.

- Go to your Settings.

- Next, tap **General**.

- Then **scroll down** to Reset.

- Tap **Reset All Settings**.

- Enter your Passcode.

- Reconfigure your Wi-Fi settings after resetting.

This would solve the problem.

Recover a Stolen iPhone SE

There is a feature on iPad Pro that can enable you to find your lost or stolen device. The App is to Find my iPhone. You can install this App from your app store if you do not have it installed already. An alternative way is to turn on your Bluetooth on the missing iPhone SE. If it goes missing, its location can be tracked easily through any nearby Apple device, even when it is switched off. If the lost device has been taken to another country, and someone walks in with an Apple device, the missing device will display its location on the nearby Apple device, which securely and anonymously send the information back to Apple and then to you.

To turn on the Find my iPhone feature:

- Go to Settings

- Scroll down and tap **iCloud** to display a menu on the right.

- Locate the **Find My iPhone** feature and toggle it On.

To Find Your Stolen Device, Follow These Steps:

- Visit www.icloud.com on any PC or any other device that has access to the internet.

- Next, log in to your iCloud account

- Click on **Find my iPhone** icon, which is also the same for any of your other devices that are connected to your iCloud account.

- Look to the top of your screen, where you will find a drop-down menu with the tag **All Devices**. Tap it to see all of the devices you have connected to your iCloud account and tap any device you wish to find.

- Select the stolen iPhone from the list of devices to be located.

- You will see a green button on the map indicating your missing iPhone's specific location.

- When you zoom in, the location of the stolen device becomes clearer.

- If you can trace it to somewhere where you misplaced it, and it wasn't stolen, let's say it is around your house, you can click on the **Play Sound** button, which will activate a sound on the missing device. This will enable you to locate it quickly.

- If you eventually cannot find the device because it is not connected to the Wi-Fi, select **Lost Mode** or **Mark as Lost**. This will remotely lock the iPhone SE. You may decide to select **Erase iPhone** to ensure that no one else can have access to your wallet or files.

- You will be asked for a phone number to be reached when you click on Lost Mode. If you didn't use a passcode on your missing iPhone, you would be prompted to enter a passcode that will lock the iPhone.

- After you have entered the phone number, click **Next**. This will open a message box that will allow you to enter a message that will display your missing iPhone's phone number. The default message is "This iPhone has been lost."

- Please call me. You can decide to edit it to whatever message you want. The iPhone will be useless to anyone who has it unless the Passcode is known. When you finish typing the message, click **Done**.

- You can also change the map on display to the standard satellite or Hybrid mode at the bottom left side of the screen.

Tips and Tricks for iPhone SE 2020

1. Save Photos and Videos

Photos and videos are vital to us. We live the memories with these precious pieces of our past. The new iPhone SE offers 4K video recording, which means better video quality with more space. 4K takes up more storage space than regular videos and therefore takes up more storage space.

You can use Google Photos to save free and unlimited compressed photos. If you want to use iCloud, you can turn on iCloud backup to save your photos there. The free iCloud account has 5GB of free storage. After that, you will have to pay for more photos and videos. The first thing to do on your iPhone SE is to turn on the photo and video backup.

2. Wireless Charging

With wireless charging, you can easily charge your iPhone wirelessly without using the charging cable. Thanks to the glass back.

You can set up a charging access point for your iPhone. To do this, make sure you find a perfect spot that you are most likely to be. For me, this is my desktop computer and next to my bed, so my wireless chargers are installed in these places. This way, I can easily charge my phone without worrying about plugging in a cable overnight.

3. Assistant Key

The physical home button is no longer on the iPhone. But the iPhone 7 brought the touch button home, it doesn't click like the physical button, but it wants it. If you are not a fan of the physical home button or, for some reason, the home button is not working properly, you can use the assisted touch feature.

Assistive Touch allows you to activate a virtual home button that can be displayed on the screen. It has more functions than the physical button. This Help button allows you to take a screenshot, change the volume, open Control Center, access Siri, rotate the screen, lock the screen, mute the sound, restart your iPhone, and more.

4. Enable/Disable 4K video

If you don't care about video quality and want to stick to 720p or 1080p videos, turn off 4K videos on your new iPhone SE. Disabling 4K in Settings will reduce video usage and save storage space. You can turn off 4K video recording on your iPhone SE in the settings.

5. Activate Dark Mode

The best solution for your eyes is the dark mode. It not only protects your eyes but also helps you sleep better. The dark mode is one of the best things that has ever happened to smartphones. You can enable dark mode in iOS 13 in **Settings> Display & Brightness> Dark**.

6. Use Third-Party Keyboards

IOS now supports third-party keyboards, so you can get a keyboard of your choice to type in. With the new iPhone SE, you can download a new iOS keyboard from the App Store and configure it based on the theme. If you like the standard iOS keyboard, you can keep it. However, I recommend using Gboard as this is the best keyboard that I think is available on any platform.

7. Use Multiple Fingerprints for Touch ID

The iPhone SE does not have facial recognition. It is the only iPhone in the current Apple range that has a physical home button. Did you know that you can set up to five fingers to set up Touch ID? Otherwise, you know now. You can tap **Settings> ID & passcode> Open fingerprint** and add up to five fingers. You can unlock your iPhone with your fingers. You can also configure the thumb if you like that kind of experience.

8. Do Not Close Applications

Closing apps does not reduce battery life. IOS is optimized to manage background applications while saving battery. This is Apple's official statement. Closing the latest apps won't help you save battery, but it will cost you more energy. If you remove an app from **Recently used apps** that you frequently use, the next time you open the app, it will consume more power, which will result in more battery drain. It is therefore advisable not to close running applications.

9. Haptic Touch

The iPhone SE doesn't have 3D Touch. It comes with Haptic Touch, which mimics 3D Touch but doesn't feel like it. You can touch and hold, and the apps will display menus depending on their features and developers. You can activate Haptic Touch in the settings.

10. Charge Your iPhone SE Faster

To charge your phone faster, use a quick charger or turn on airplane mode and charge it. Enabling airplane mode will disable all other connectivity options and use less power. You can also turn off your iPhone and charge it for faster charging.

11. Tip for Saving the Calculator

The iOS calculator was a sworn enemy until I got practical advice. It was boring not knowing how to erase numbers. Now I know that swiping the numbers left or right will delete an entry from the calculator. Instead of deleting all entries, you can scan integers, and one will be deleted.

12. Cancel the Type By Shaking It

Are you writing a long paragraph or story and want to quickly delete the sentences you entered? Undoing by shaking is your favorite function. You can turn on the **Shake to Undo** feature in **Settings> Accessibility> Undo**.

13. Utilize Camera Filters to Take Better Selfies

Selfies are essential if you have an iPhone. The iPhone SE 2020 has an excellent camera that supports portrait mode thanks to the A13 Bionic chipset. You can improve it further by using the filters in the standard camera app. I love the Vivid Filters for the iPhone. This makes the images more saturated. You can likewise choose other options, such as Vivid Warm, Vivid Cool, Dramatic, Dramatic Warm, Dramatic Cool, Mono, Silvertone, and Black.

14. Transfer Data Directly From the Old iPhone

Are you from an old iPhone and upgrading to the new iPhone—iPhone SE 2020? There is no need to use iTunes or iCloud to transfer data between the old iPhone and iPhone SE 2020. Instead, keep the old iPhone closed while setting up your iPhone SE, and you will automatically be given the option to transfer all data directly from it.

15. QuickTake

The iPhone SE 2020 includes QuickTake in the camera app. Similar to the iPhone 11 and iPhone 11 Pro series, you can long-press the shutter button to immediately start recording a video, just like you did with Snapchat. The video stays in the same setting as the photo, which is impressive.

If you want to record videos longer, you can swipe right on the shutter button to lock it in video recording mode.

16. Try Using a Gesture

With iOS 13, Apple finally added support for gesture input to the standard iOS keyboard. Gesture input is even more practical on the small 4.7-inch screen of the iPhone SE. The feature works the same as SwiftKey and Gboard.

17. Automate Tasks Using Links

Your iPhone SE is pretty smart. With the new shortcut app, you can automate repetitive tasks or combine multiple tasks. The Shortcuts app lets you create a shortcut to turn off Do Not Disturb, text someone, read the latest headlines, and turn on the lights with a single command.

18. Vertical Lighting

Although it has a single rear camera, the iPhone SE allows you to take vertical photos. It also has a portrait lighting feature that allows you to change the background of a portrait photo. The Stage Light effect lets you switch between Studio Light and Contour Light on a completely black background, and it looks really good. Please note that portrait mode on the iPhone SE 2020 does not work with pets and other objects.

Best of all, you can preview all of these effects live before you hit the trigger. Switch to portrait mode, and you'll see a carousel below. Use this option to switch between the various portrait lighting modes.

19. Easily Cut, Copy, and Paste Text

It's more of an iOS 13 feature, but it's still worth a mention. There are new gestures in iOS that make it easy to cut, copy, and paste the text. Use the three-finger pinch gesture to copy text, a three-finger pinch gesture to cut the text, and a three-finger zoom gesture to paste the text.

20. Connect Multiple Airpods to Your iPhone

The iPhone SE 2020 supports audio sharing, which allows you to share any audio you want to play on your AirPods with another pair of AirPods too. Just open the AirPlay section in the Control Center and open the second pair of AirPods to connect and use with your AirPods.

21. Optimize Battery Charge

If you plan to use your iPhone SE in the next few years, you will need to enable the battery charge feature on your iPhone SE. In this way, you realize what personal corresponds to the correct use and charging of your iPhone. It will be closed or even heard until it is over 80% loaded by the time you listen to it. This helps to fulfill the responsibility of the battery.

22. You Must Have the Sensations of Home Tastes

The home button on the iPhone SE 2020 doesn't actually have a particular taste. This is why the Force Touch trackpad works on MacBook. The good thing is that you can sense his comments and your tastes. Go to settings. Then general home and looked for the listened option.

23. Silence Unknown Caller

All of you can have the following iPhone rights in the Contacts app. Go to **Settings> Messages**, and you will see the **Silent Unknown Events** option. These voices are then damaged directly to your voicemail when you use them.

24. Express Card With Power Reserve

The 2020 iPhone SE is tested with the Express Card Control with power reserve. The feeling of power reserve is new and the same that you can pay for the ride on the subway with your iPhone even if the battery runs out quickly. The function works with all public transport or payment cards.

25. Look for Tabs for Safari and More

If Safari has a lot of tabs listed and you have personal control, this tip is very important to you. You have to tap on the tab icon and then proceed to finish. One such list is heard above.

Enter the keywords you are looking for, and they will be the tabs the keyword belongs to. You can also long click on the **correct Cancel button**. An option will appear to protect the cards that match your search terms.

Reopen closed Safari tabs:

If you have someone else, you can keep the "+" flavor to keep a list with some of the full tabs.

Self-closing Safari tabs:

If you have Safari, you might not close the tabs but quit Safari itself. This becomes one of the junk cards over time. To fix this, you can flatten Safari so that the tabs are closed.

Go to **Settings> Safari> Close Tabs**. There are three options to choose from. Always know the cards. When you tap the selection, Safari will have the tabs for you.

Add all tabs:

Just like reading all tabs in Safari, it has its trick; all tabs will be safe. If you remove and hold the bookmark icon, you will get more options. Now the required bookmark for X tabs. It belongs to a Safari folder and belongs to all currently open keys with a bookmark. You can still hear all the tabs.

26. Consider Apple Care Warranty and Status in Settings.

Apple Care Warranty and Visual Inspection belongs online; please visit the Apple website and register. In iOS 12.2, Apple made it easy to check the status directly on the iPhone in the settings application.

You just have to go to **Settings> General> About**. You will find two new sections. An almost perfect guarantee and apple care fall off if you have one. Here you can easily change your status.

27. Quickly View Screenshots

Anyone who uses the iPhone knows how to take a screenshot (if you are new, hold the power button, and listen to the volume to take a screenshot). One of the trusts to take a screenshot is the one with the managers.

When you've received a screenshot, you've tapped and decided on the share sheet icon to share or have. Then go to the Photos app and share it. You can keep the small preview image open from the share sheet.

28. More Precise Settings

This tip is true if you have the schedule at the right time. This can become the application timer only every 5 seconds. Transparent timer sequence (5, 10, ranges from 15 seconds) the option is set to 1 second right to 5 seconds.

A double-tap will return to the wheel for 5 seconds. This tip is very useful when you want to set the approximate time.

29. Find Similar Songs on Apple Music

Many people probably know this tip. Finding the same songs on Apple Music is as easy as going unnoticed.

You can just tap a 3D song and tap Create Station. This will create a station with multiple songs similar to the ones you touched in 3D.

30. Strengthening Stakeholders

This simple setup makes the speakers louder than before. Tap **Settings> Music> EQ**, find and tap **Late night**. This equalization setting increases the loudspeaker volume. This is a simple trick to turn up the volume on your speakers without damaging them.

31. Quickly Send Your Current Location to iMessage

Sharing a location is very important to most users daily. You can easily share your location by clicking **I'm at**. The Current Location tab automatically appears in the dictionary area. Tap your current location to instantly share your location with your friend. It's currently a quick way to share your location.

32. Various Pronunciations From Siri

Did you know Siri can pronounce your name differently than the standard pronunciation? It's very easy. You can just ask him to say your name however you like.

For example:

- James — "Siri, what's my name?"

- Siri — "James."

- James — "It's not pronounced right."

- Siri — Could you do it again?

- James — "James."

- Siri — "Okay, thanks. What pronunciation should I use?"

A pronunciation is now displayed. You can listen to them and choose the one that suits you best.

33. File Notification

If you have a lot of apps and notifications in one folder, this tip is helpful. You really don't want to tap the folder and swipe left or right to see the notification for each app. You can just tap the folder in 3D, and the notification badge app will be listed. This is useful for people who use a lot of files.

34. Reduce the Brightness to Below 0%

This trick allows you to lower the brightness and make the screen darker than Apple allows. This is useful when you are in deep darkness. All you have to do is go to **Settings> General> Accessibility> Show Accommodation** and press **Reduce White Point**.

35. Ping the Lost Device With a 3D Key

Pinging a device is useful if you've lost the device. Usually, you can ping a device by going to the Find My iPhone application. The 3D Find My iPhone button displays a list of the devices that have connected to your iPhone. Tap the lost device to ping it immediately.

36. Back Up Your Data By Deleting It

Did you know that if an incorrect password is entered ten times in a row, the iPhone feature can automatically delete its data? Sounds good, doesn't it? Do the following:

- Configure the Settings app on your iPhone.

- Scroll down and open face recognition and passcode.

- Enter your password.

- Activate data deletion.

37. Set the Timer for the Music

Don't you fall asleep before you turn off the music? Here's how your watch can set a timer to turn off everything you see or hear on your iPhone and iPad:

- Start the clock app on your iPhone.

- Touch the **Timer tab** at the bottom of the screen.

- Select the **duration.**

- Touch **End timer**.

- Scroll down and go to **Stop Reading**.

- Touch **Setup** at the top right of the screen.

38. Get Rid of Annoying Ads

We do not allow disclosure of your personal information as part of your story. Go to Settings. Then to **Privacy>Advertising**. Then, turn on limited ad tracking.

39. Automatically Close the Browser Tabs

Would you like to automatically close the Safari browser tabs to make your browser work more efficiently? Follow these steps:

- Configure the Settings app on your iPhone.

- Scroll down, and it will show up on Safari.

- Touch **Close Tabs.**

It is possible to configure the closing of the tabs after a day, a week, a month, or to exit the manual mode.

40. Start a Group Call in Real-Time

You've probably done FaceTime. Thousands of times with your friends and family, but have you seen it multiple times? Here's a comment:

- On your iPhone, set up the **FaceTime** application.

- In the top right corner of the screen, press **the + button.**

- Enter the name or number of the person you want to call first.

- Enter up to 30 additional contacts.

- Press **Video or Audio** to make the FaceTime call.

41. Personalize Text Messages When You Ignore a Call

We know that you are busy and that sometimes you want to ignore a call. But instead of leaving your friends or family on the ground, you can send a written message. Sometimes you don't have time for it.

Follow the steps below if you want to customize 3 more options:

- Tap the Settings app.

- Scroll down and tap **Phone**.

- Tap **Reply with the text.**

- Choose the area you want to change

- Enter the new text.

42. Add Multiple Faces to Face ID

Here's what you can do if you want to register multiple faces to unlock your phone

- Tap the Settings app.

- Click **Face ID & Passcode** as it scrolls down.

- Enter your password.

- Touch **Set up another appearance.**

- Scan your face and follow the on-screen instructions.

43. Measure Objects With Your iPhone

It seems that whenever you need to measure something, you can never find the meter to do it. Fortunately, you can use your iPhone's built-in measurement tool to measure parts and objects. It's not entirely accurate, but it's a useful tool when you need to measure an object or the distance between two things at the same time. Follow these steps:

- Launch the measurement app.

- Move your iPhone so that the device can scan the area.

- Hold your iPhone so that the camera points towards the object you want to measure.

- Move your iPhone until you see a white circle with a dot in the center.

- Align the white point with the edge of the item you want to measure.

- Press and hold the **white button** with the **+ sign**.

- Scroll to the opposite edge of the item.

- Hold down the **white button** with the **+ sign**.

- The application will display the approximate measurement.

44. Limit iPhone Use to Screen Duration

Guilty of using your iPhone too often? There's an easy way to cut down on time spent on social media, shopping online, or scanning titles with Screen Time. Here are the instructions on how to enable the clock on the screen:

- Tap the Settings app.

- Now tap **Screen Time.**

- Then tap **Application Limits**.

- Then tap **Add limit**.

- Touch a category.

- Tap **Add**.

- Choose the time.

- Hold down the top left arrow to save.

45. One-Handed Keyboard

IPhone screens are getting bigger and making typing difficult, especially with one hand. Fortunately, you can enable the one-handed keyboard feature on your iPhone so that you can type with one hand and do whatever else you need to do with the other. Follow these steps:

- Press and hold the **emoji or globe icon** in the lower-left corner of the keyboard.

- Choose the **left or right keyboard symbol**.

- Press the **arrow** in the area created by the keyboard offset to return to normal.

46. Cut the Thread of Text Messages

Do you receive several short messages from a friend? Here are some great tips for you. You can hide SMS alerts so you won't be bothered by alerts when you receive SMS from a specific person or group. Follow these steps:

- Launch the **messaging app**.

- Swipe left in the text chat you want to disable.

- Click the **Hide Alerts** button.

47. Enter Siri

Talking to Siri in public is a little crazy. Fortunately, you can use your finger to ask Siri questions instead of your voice. Follow the steps below:

- Tap the Settings app.

- Now hit **Accessibility**.

- Scroll down and tap **Siri**.

- Toggle on **Type to Siri**.

Great tip: if you ever lose your iPhone at home, just call "Hey Siri," and it should beep. Continue until you find your phone.

Also, make sure your phone is in silent mode. Otherwise, Siri will share your answer with everyone.

48. Use The Document Scanner in the Notes Application

Do you want to quickly scan a document directly from the Notes application? Follow these steps:

- Start the **Notes application.**

- Now start a new note or open an existing one.

- Then press the **+ icon** in the center above your iPhone keyboard.

- Then tap on **Scan Documents**.

- Use the shutter button or one of the volume buttons on your iPhone to take a photo of your document.

- Adjust the corners of the document by touching and dragging as needed.

- Finally, tap on **Save**.

49. Quit Applications That Ask for Feedback

Turn off those pesky app ratings and check pop-ups on your iPhone. **Go to Settings> tap iTunes and the App Store> Turn off app ratings and ratings.**

50. Change Screens

Did you know you can use your iPhone to edit your screenshots before saving or sharing them? Follow these steps:

- Take a screenshot by pressing the side button and volume up button at the same time.

- Quickly tap the screenshot thumbnail at the bottom left of the screen.

- Now you can crop, draw, highlight parts, etc.

- When you're done editing, tap **Done** in the top left or the share icon in the top right to send it to a friend or family member.

51. Download Unused Applications

You've likely received an "almost full memory" pop-up multiple times, and you may even have deleted some apps and photos to free up space to install iOS 13, especially if you're using a 16GB device. If you're one of those people, this little tip will save you a lot of stress and space over time. Automatically remove hidden apps without deleting their documents and data. To download apps you don't use and save valuable storage space, do the following:

- Tap the **Settings app**.

- Go down and click **iTunes and App Store**.

- Activate downloaded applications that are not being used.

- Deleted applications are turned gray out on the home screen and can be reinstalled with a tap of the finger.

52. Customize the Control Center

Did you know that your Control Center parameters are fully customizable? You can easily set the parameters and functions that you frequently use where you want. Follow these steps:

- Tap the **Settings** app.

- Now press the **Control Center**.

- Then press **Customize Controls**.

- Then tap the **green "+" symbol** next to an item you want to add and the **red "-" symbol** to remove the items you want to remove from the Control Center.

- Touch, hold, and drag the **three-bar icon** to reorder the orders.

- When you're done, tap the **Back button** in the top left corner of the screen.

- Finally, swipe up from the bottom of the screen to reveal your custom control center.

53. Do Not Disturb While Driving

This feature is designed to prevent accidents and reduce distractions while driving. Try not to Disturb blocks approaching calls, instant messages, and different notices when your iPhone distinguishes increasing speed or interfaces with your vehicle's Bluetooth.

When someone tries to contact you, your iPhone will send an autoresponder message to let them know that you are driving. Adhere to the below to activate Do Not Disturb while driving.

- Tap the **Settings** app.

- Now tap on **Do Not Disturb**.

- Under "Do not disturb while driving," automatically tap if you want to activate the setting when your iPhone detects movement.

- Touch When connected to Bluetooth for the car if the car is compatible with Bluetooth.

- If you do not want the function to be activated automatically, press manually and then add **Do not disturb while driving** in the Control Center (see previous tip above).

- Tap **Auto-Reply** to customize and edit the message if you think it is necessary.

- Touch **Auto Reply** if you want to change who receives the auto-reply message. You can select **All Contacts, Favorites, Recent Contacts,** or **None.**

54. Replacement of the Text

Your iPhone has a text replacement feature that allows you to type a few characters instead of typing a sentence or entire sentence forever. Follow the steps below to set up the text replacement feature on your iPhone:

- Tap the **Settings app**.

- Now press **General**.

- Then press **Keyboard**.

- Then tap **Replace Text**.

- Click on **the + symbol** in the upper right corner.

- In the Phrase field, enter the word or phrase you want to link to.

- In the Link field, enter the text you want to replace with the phrase.

- Then tap **Save** in the top right corner.

Whenever you tap the link in a text box, your iPhone replaces it by pressing the spacebar after pressing the spacebar.

55. Define the Vibration Settings

You may already have custom ringtones for certain calls, but that doesn't help if your phone is in silent mode. Fortunately, you can create specific contacts and assign custom vibrations. Follow these steps:

- Tap the **Settings** app.

- Tap **Sounds & Haptics**.

- Touch the ringtone.

- Press **Vibration**.

- Select to create a new vibration.

- Create a new vibration by holding it down as desired.

- Press and hold the stop button in the lower right corner.

- Click **Save** in the upper right corner.

- Finally, name your vibration.

Once you've created your custom vibration, simply do the following to assign it to a contact:

- Launch the **Contacts app**.

- Select the contact you want to assign the custom vibration to.

- Press the **Edit** button in the upper right corner.

- Select vibration.

- Select the new vibration created earlier.

56. Use Your iPhone as a Spirit Level

Did you know that your iPhone can be used as a spirit level? Well, there is a tool in your iPhone that you may not know about: a spirit level. Follow these steps to access it:

- Launch the **measurement** app.

- Press **Level** in the lower right corner.

- Place the phone flat on the surface you want to check for flatness.

You can measure the difference between the two surfaces by touching the screen. The red border shows how much the two angles vary.

57. Use the Headphone Cable to Take a Photo

Shaking hands don't give you good photos? Well, you can take a snapshot with the volume up or down buttons on the headphones.

58. Hide Photos

At times, you have photographs in your camera roll that nobody would be pleased with. Fortunately, you can prevent this from happening by hiding some photos. Follow these steps:

- Launch the **photo app**.

- Touch the album that contains the photos you want to hide.

- Tap the **Select** button at the top right.

- Select the photos you want to hide.

- Tap the **Share** button in the lower-left corner.

- Tap the **Hide** button in the lower right corner.

- Tap **Hide Photos**.

The following tips will help you get to know your iPhone SE better, discover some of its hidden features, and so much more.

Restrictions of the New iPhone SE

1. Doesn't Have the Latest Design

The iPhone SE 2020 does not have the latest designs from the latest iPhones. The design is very similar to that of the iPhone 8. The cheap iPhone, the end-to-end screen design and FaceID, the iPhone XR 2018 with TouchID, and huge bezels above and below.

2. Due to an Older LCD Screen Than a Two-Year-Old iPhone

Another reason for disputes is the screen used by the iPhone SE 2020. The smartphone is equipped with a 4.7-inch Retina HD display and resolutions of 1334 x 750 pixels. It's the lowest screen resolution, even compared to iPhone standards, in the past three years.

3. Still a Single Rear View Camera for $399

Another hardware limitation is the only rearview camera that costs Rs 42,500 on a smartphone. At this price, you can, of course, get smartphones with three or four reversing cameras, which have more functions and special lenses for various activities. Let's not forget that the rearview camera of the iPhone SE 2020 has a 12-megapixel sensor and has less than half of the Android phones at this price.

4. No Connection for Headphones 3.5 Mm

Apple lowered the connection for headphones 3.5 mm iPhone cheaper in 2020. If the world is someone who comes from an Android phone, they missed it, even if you move quickly towards wireless headphones. The idea is good to be in line with Apple's philosophy of wireless communication, but one may wonder what could prevent it from being brought to its cheapest smartphone, which is likely to work well in developing markets. This is also the market where people are still switching from wired to wireless headsets.

5. Unfortunately, the Same Battery Capacity as iPhone 8

Has a battery capacity of less than 3000 mAh, and it is surpassed by Android phones. As Apple noted on the specification page, it uses a battery like the iPhone 8. Since the launch of the iPhone 8, if dismantling of the device, it turns out that approximately contains an 1800 mAh battery. This may not be impressive for someone looking at the specifications.

Compared to iPhone 11

The price for iPhone SE is $ 399 and $ 699 for iPhone 11, and iPhone SE has a smaller screen size of 4.7, and iPhone 11 is 6.1. The iPhone SE has a single rear camera compared to the two rear cameras of the iPhone 11, one of which is an ultra-wide camera.

Another important difference between the two phones is the security of the phone, which is how to unlock the device. While the iPhone SE uses a Touch ID sensor, the iPhone 11 has a face ID. We generally prefer the latter, but Touch ID works well and very quickly.

Both phones are the Bionic A13- chip processor that supplies power and is charged through wireless charging.

Compared to iPhone 8

The iPhone SE for $ 399 replaces the Apple Series iPhone 8 for $449, and you get a better phone for less. There are some important differences between the iPhone 8 and iPhone SE, although they have a similar design. The first is the A13 Bionic processor, which is more powerful than the A11 Bionic chip in the iPhone 8.

The new iPhone SE has a better camera function. The cameras rear 12MP and front 7MP contain a similar resolution, but computer photographic enhancements of chips Aion Bionic iPhone would do better. This includes portrait mode and vertical lighting effects for the rear and front cameras.

While it is predicted that the battery capacity of the iPhone SE will be similar to that of the iPhone 8s, the A13 chip in the new iPhone SE can give the new Apple phone a longer lifespan.

Conclusion

The iPhone SE is an amazing device. It is my concern to teach you how to use the smartphone in an easy and understandable way without bluffing, and I hope you're satisfied with my level of input. I made this for you, and presumable you can now do all configurations with your iPhone SE (2020). I hope you find this guide useful and insightful, and it has helped you to find solutions to the most important features you ever wanted. Good luck and cheers.

CPSIA information can be obtained
at www.ICGtesting.com
Printed in the USA
LVHW060052210221
679536LV00028B/659